Ultra-Dense Heterogeneous Networks

Driven by the ever-increasing amount of mobile data, cellular networks evolve from small cell network to ultra-dense heterogeneous networks, to provide high system capacity and spectrum efficiency. By bringing base stations (BSs) to similar number magnitude of served users, ultra-dense heterogeneous networks would definitely bring unprecedented paradigm changes to the network design. Firstly, along with densification of small cells, inter-cell interference becomes severe and may deteriorate performance of mobile users. Assigning network resources including bandwidth and time slots, while avoiding interference, desires serious consideration. Secondly, the coverage area of BSs becomes small and irregular, resulting in much frequent and complicated handovers when mobile users move around. How to ensure continuous communication and implement effective mobility management, inter-cell resource allocation and cooperation, remains a challenging issue. Thirdly, such dynamic change in spatial dimension motivates us to re-investigate available and ongoing communications and networking techniques, such as massive MIMO, CoMP, millimeter waves (mmWaves), carrier aggregation, full duplex radio, and D2D communications.

To address the aforementioned challenging research issues, this book will investigate the service and QoE provisioning in ultra-dense heterogeneous networks. In particular, firstly we introduce ultra-dense heterogeneous networks by careful definition regarding spatial deployment, generic characteristics, and requirements of ultra-dense heterogeneous networks in order to ensure QoE of mobile users. Secondly, we depict the resource management among small cells in close proximity, mobility management for mobile users (addressing the super-frequent handovers), and interference management (dealing with the interference due to frequency-reuse in the vicinity). Thirdly, we study the enabling factors, and the integration of ultra-dense heterogeneous networks with enabling technologies, such as massive-MIMO, cloud-RAN, mmWaves, D2D, IoT. Finally, we conclude the book and indicate future directions and challenges.

Ultra-Dense Heterogeneous Networks

Wen Sun

Qubeijian Wang

Nan Zhao

Haibin Zhang

Chao Shen

CRC Press
Taylor & Francis Group
Boca Raton London New York

CRC Press is an imprint of the
Taylor & Francis Group, an **informa** business

First edition published 2023
by CRC Press
6000 Broken Sound Parkway NW, Suite 300, Boca Raton, FL 33487-2742

and by CRC Press
4 Park Square, Milton Park, Abingdon, Oxon, OX14 4RN

CRC Press is an imprint of Taylor & Francis Group, LLC

© 2023 Wen Sun, Qubeijian Wang, Nan Zhao, Haibin Zhang and Chao Shen

Library of Congress Cataloging-in-Publication Data
A catalogue number has been requested for this title.

ISBN: 978-0-367-70950-1 (hbk)
ISBN: 978-0-367-70951-8 (pbk)
ISBN: 978-1-003-14865-4 (ebk)

DOI: 10.1201/9781003148654

Typeset in Nimbus
by KnowledgeWorks Global Ltd.

Publisher's note: This book has been prepared from camera-ready copy provided by the authors.

Contents

1 Introduction

1.1 PROSPECTS AND REVIEWS OF COMMUNICATION NETWORKS

According to the forecast of Cisco Global Cloud Index, there will be more than 847 ZB data generated each year, of which more than 50% of the data needs to be analyzed, processed, and stored at the edge of the network [1]. Meanwhile, worldwide online devices would exceed 28 billion and generate up to 4.8 ZB of traffic per year [2]. The transmission and analysis of such massive data poses unprecedented challenges on the current telecommunication system.

Nowadays, to meet the rapidly rising data demand, researchers have started to conceptualize 6G with the vision of connecting everything, transmission over millimeter wave (mmWave) and THz, and integrating sensing, communication, computation, and control functionalities. Several companies (Nokia, Ericsson, Huawei, Samsung, LG, Apple, Xiaomi), as well as several countries (Finland, British, Germany, and China), have shown interest in 6G networks. In 2018, Finland took the lead in announcing research plans to develop a complete 6G ecosystem, which will last for 8 years with a total amount of $290 million [3]. After that, Britain and Germany began investing in projects including quantum technology. On November 20, 2019, China announced the launch of 6G research and began to develop spectrum technologies such as terahertz and mmWave [4]. Although the development of 6G is still in its infancy, 6G networks are expected to provide extremely high data rate and support applications beyond current mobile use scenarios, such as virtual reality (VR), augmented reality (AR), and Metaverse.

Spectrum extension and spectrum efficiency improvement (e.g., cell splitting and densification) have been considered the most effective means to deliver ever-increasing data capacity and improving quality of user experience. Spectrum extension aims to achieve high data rate through the extension of unused electromagnetic spectrum, e.g., mmWave, Terahertz. Whereas, the efficiency of spectrum can be effectively improved using emerging techniques, such as Massive-MIMO and ultra-dense networks.

Spectrum Extension Technology

As shown in Fig. 1.1, electromagnetic spectrum can be categorized into radio, microwave, infrared, visible light, ultraviolet, X-Ray, and Gamma-Ray. Wireless communications typically exploit the radio bands and the microwave bands within 6 GHz. 5G extended the usage of spectrum resources to the range of 24 GHz and 100 GHz to increase communication capacity. However, the expanded frequency band are still insufficient to meet the growing bandwidth demand. To further support bandwidth-hungry and latency-intolerant applications, terahertz transmission becomes a complementary wireless technology for Beyond 5G and 6G.

DOI: 10.1201/9781003148654-1

1

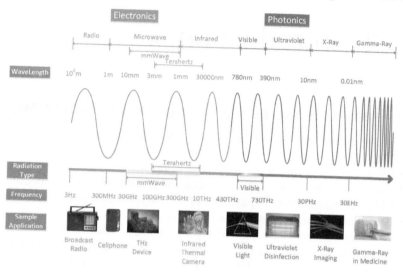

Figure 1.1 Spectrum map including mmWave and terahertz frequency bands

- Terahertz communication
  Terahertz band, also known as submillimeter band, consists of the spectral band within the frequencies from 0.3 to 3 THz. Terahertz communication can provide tens of GHz of available bands for high-speed data transmission (exceeding several hundreds of gigabits fer second). Furthermore, terahertz communication has high directionality and short pulse time, which can reduce interference while ensuring transmission security. Terahertz transmission has become a wireless backhaul extension of the optical fibers for scenarios including hotspots and last-meter indoor wireless access. The latter is either infeasible or very costly for transmission in rural or remote region. However, the terahertz wave can be easily absorbed in the air due to its poor diffractive ability. In the future, Terahertz communication will inevitably become the feasible backhauling technology for ultra-dense networks [5].
- Visible light communication (VLC)
  Visible light communication, as a subset of optical wireless communications, refers to a data communication using visible light between 400 and 800 THz. The technology enables transmission at 10 $Kbit/s$ using ordinary lamps, or beyond 500 $Mbit/s$ using micro-LED. It is believed that with the development of VLC and LED technology, the transmission rate of VLC will reach Tbit-level per second [6]. Compared with traditional radio communications, VLC is feasible for ubiquitous communication medium, as lighting devices are used everywhere, e.g., indoor/outdoor lamps, TVs, car headlights, and traffic signs. VLC can provide a large amount of potentially available spectrum resources without the authorization of the spectrum regulator. Moreover, VLC is insusceptible to external electromagnetic

interference while generating no electromagnetic radiation. Since the transmission range of visible light is limited by the structure of buildings, VLC can reduce inter-cell interference and ensure reliable transmission.

Spectral Efficiency Improvement Technology

- Massive-MIMO
  Massive-MIMO is an extension of MIMO that groups antennas together at the transmitter and receiver for improved throughput and spectrum efficiency. In 5G, the 8-antennas MIMO system evolves toward the massive-MIMO system with $256-1024$ antennas, which rapidly increase the capacity and spectral efficiency of the system. It is expected that the number of antennas deployed by the MIMO system will exceed $10,000$ in 6G systems. A large number of antennas will bring the following advantages: (1) by applying spatial multiplexing technology to transmit hundreds of parallel data streams on the same channel, the system can significantly increase energy efficiency and reduce latency; (2) by utilizing hundreds of available beams, the system can simultaneously serve multiple UEs to significantly increase network throughput; (3) by forming ultra-narrow beams to overcome the propagation loss of mmWave and the terahertz wave, the system can greatly reduce inter-cell interference. Despite the promising advantages, complex algorithms are desired to find the exact location of UEs so as to apply accurate beamforming technology. Beam management and beam steering are key technologies to enable massive-MIMO to be applied in practical scenarios [7].

- UDN
  By bringing base stations (BSs) to the approximate spatial scale and number magnitude of users, ultra-dense networks (UDNs) would definitely bring unprecedented paradigm changes to the network design. Firstly, along with densification of small cells, inter-cell interference becomes severe and may deteriorate performance of mobile users. Assigning network resources including bandwidth and time slots, while avoiding interference, desires serious consideration. Secondly, the coverage area of BSs becomes small and irregular, resulting in much frequent and complicated handovers when mobile users move around. How to ensure continuous communication and implement effective mobility management, inter-cell resource allocation and cooperation, remains a challenging issue. Thirdly, such dynamic change in spatial dimension needs us to re-investigate available and ongoing communications and networking techniques, such as massive MIMO, CoMP, millimeter waves (mmWaves), carrier aggregation, full duplex radio, and D2D communications [8].

1.1.1 FRONT END NETWORK

The front end network generally refers to the network that provides access for UEs. The rapidly increasing data demand urges the front end network or access network

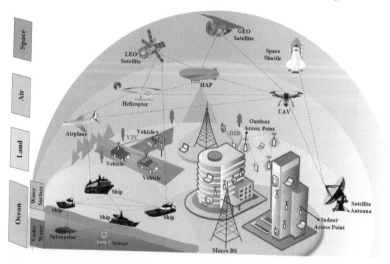

Figure 1.2 The four-layer architecture of the future network (the ocean layer includes both the underwater and the water surface)

to evolve in both the breadth and depth of coverage area. In particular, densification of terrestrial base stations becomes a general tendency to increase network capacity. As shown in Fig. 1.2, the future 6G networks are expected to provide ubiquitous and reliable service through the Space-Air-Ground-Ocean Integrated Network, integrating satellite systems, aerial networks, terrestrial networks, and ocean communications. Benefiting from their wide coverage, high flexibility, and strong resilience, the Space-Air-Ground-Ocean Integrated Network are competent for seamless wireless communication services in emergency situations.

Space Network

Space network, first initiated by NASA, aims at combining space and ground elements to support spacecraft communications in Earth vicinity, i.e., supporting communication services for satellites, space shuttles, and space robots [9], and further provide satellite Internet access to most of the Earth. Nowadays, the satellite network plays an essential role in communication services such as voice calling and video broadcasting. Nowadays, several satellite constellations are moving towards commercialization [10]. Starlink, operated by SpaceX, is a satellite internet constellation providing satellite Internet access to most of the terrestrial users. The constellation has grown to over 1700 satellites through 2021, arrange another 30,000 additional satellites, and will eventually consist of many thousands of mass-produced small satellites in low Earth orbit (LEO), which communicate with designated ground transceivers. By November 2021, the beta service offering is available in 21 countries. Moreover, on February 27, 2019, OneWeb authorized by UK and the Federal Communications Commission (FCC) launched six satellites into the expected orbit. The success of the launch laid the foundation for subsequent satellite deployment, OneWeb plans to eventually build a satellite constellation consisting of 720 LEO satellites. China Aerospace Science and Technology Corporation (CASC) will also launch 9 LEO satellites as a pilot demonstration of the Hongyan system. The Hongyan system will eventually include 320 satellites and is expected to be completed by 2025 [4]. While non-geostationary orbit (NGSO) satellite system faces challenges in the comprehensive deployment and the integration with mobile wireless network, LEO satellite system has found its applications in real scenarios. It is expected that in the near future, the satellite communication system will provide UEs with uninterrupted, high-quality connections and ultra-fast network access services at data rates up to 10 Gbps.

Air Network

The air network layer provides communications for ground devices through unmanned aerial vehicles (UAVs), airplanes, and high altitude platforms (HAPs), where signals mainly work in the low frequency (i.e., microwave and millimeter wave bands). The air network can be broadly divided into two categories: (1) HAP network which operates primarily in the stratosphere; (2) low altitude platform (LAP) network which operates primarily at low altitudes not exceeding 10 kilometers. The HAP network is characterized by the wide coverage and long service lifetime. The LAP network has advantages in rapid deployment and flexible configuration, which bring opportunities for short-range communications. UAVs in the LAP network can also act as relay nodes in long-distance communications to provide continuous service for terrestrial networks. The UAV wireless network can maintain reliable communication service even if the infrastructure is severely damaged, therefore, the air network plays an important role in emergency disaster relief or search and rescue.

Terrestrial Network

The terrestrial network layer is responsible for providing communications for ground users, large-scale IoT devices, and vehicles, etc. The terrestrial network system is currently the most developed mobile communication system. However, the growing number of terminal devices has brought challenges to network management and optimization. AI-based algorithms have been developed to improve the performance of a communication system through data training and learning procedures. Besides, the terrestrial network has the following features:

- Fusion of communication, computation, and caching
  The expansion of wireless networks from single function to multiple functions including information transmission, storage, and processing can maximize the utility of networks, thereby providing high-quality services for multiple applications. The system can make the best decisions based on different resource states by integrating intelligent sensing, communication, computing, and caching. In particular, proactive caching technology can

intelligently cache data from communication nodes according to the popularity or demand rate of known nodes, which is helpful to reduce delays and power consumption in the data routing and transmission [11].
- Wide applications of multiple access technology in multiple resource domains

  In 4G networks, Orthogonal Multiple Access (OMA) technique is an effective way to achieve multiple access. However, with the development of network densification and the rapid increase in the number of UEs, the traditional OMA technique has evolved into Non-Orthogonal Multiple Access (NOMA) technique. In general, orthogonalization is achieved by resource allocation through slicing available resources in time, frequency, code, or space domains. However, NOMA technique mixed with the traditional OMA technique can divide resources jointly in the multi-dimensional space such as power, phase, and code. Moreover, NOMA can use the additional power domain to achieve multiple access [12].
- Popularization of Large-scale IoT

  IoT and AI are regarded as the core technologies for future network development. The concept of IoT is to extend network connectivity to multiple devices with capabilities of sensing, detection, and data mining and analysis. The applications of large-scale IoT devices can provide timely and effective information for the communication and management of a system. In particular, the applications of large-scale IoT devices can also provide sufficient sample information for the automated management of AI. IoT technology has extremely high development prospects in the fields of industry, agriculture, medical treatment, and education. It is believed that the technological combination of IoT and AI will greatly stimulate the potential of network systems, thus improving the quality of our lives significantly.

Ocean Network

Ocean network is responsible for providing communication services for sensors, submarines, and ships. Although human activities are mainly on land, oceans occupying a large area of the earth are inextricably linked with our life. The ocean network can be divided into two parts: the water surface network and the underwater network. The water surface network mainly realizes the communication between ships, so that passengers who travel by ship across the ocean can also enjoy network services. The underwater network enables communication between underwater equipment such as submarines and sensors mainly used in military and commercial fields.

The ocean contains abundant resources that directly determines the ecological quality of the earth. Information such as water quality, underwater organisms, water temperature, and pollutant content need to be fed back to people in a timely manner. If an underwater network is deployed, human capabilities of monitoring and management for the ocean will be greatly improved. Considering the different characteristics of underwater propagation, multiple technologies such as radio frequency (RF) communication, acoustic communication, and quantum communication can be used to achieve underwater wireless communication. Technologies such as laser communication and neutrino communication have received increasing attention as emerging technologies in underwater communication.

1.1.2 CORE NETWORK

The core network as the key part of the communication system is mainly responsible for data process and dispatching [13]. The initial core network is the fixed telephone network mainly composing circuit switching functions. Network elements, hardware equipment, and service requirements in the system have evolved over time.

- 2G: global system for mobile communications (GSM)
  The GSM [14] core network is based on digital voice transmission technology. The GSM core network mainly is responsible for the switching function of the GSM, as well as the database function in the mobility management and security management. It plays a regulatory role in the communication between GSM UEs and other communication network UEs. The GSM core network can provide UEs with a transmission rate of 10 *Kbps*.
- 2.5G: general packet radio service (GPRS)
  Since the GSM core network does not support the function of packet switching services, GPRS [15] overlay network is introduced to enable text and image transmission. The GPRS core network introduces the concept of IP address, which completely changes the traditional mode of core network that utilizing dedicated channel for circuit switching, thereby achieving the transmission and format conversion of data packets between different data networks. The GPRS core network can provide UEs with a transmission rate of 50 *Kbps*.
- 3G: universal mobile telecommunications system (UMTS)
  Compared with 2.5G, the 3G core network UMTS [16] is still composed of circuit switched domain and packet switched domain. However, the 3G core network separates the control and bearer architectures to replace the original integrated structure. In data transmission mode, UMTS adopts the IP bearer method to replace the traditional time division multiplexing (TDM) bearer mode. The UTMS core network can provide non-mobile UEs with a transmission rate of 2 *Mbps*, moreover, it allows UEs to enjoy various services such as video calls and mobile Internet access.
- 4G: evolved packet system (EPS)
  In order to meet the transmission requirements of mobile broadband data, the 4G EPS [17] core network adopts an efficient transmission architecture based on IP services. 4G core network is built on top of the 3G packet switched domain. By using the packet gateway to connect to the 3rd generation partnership project (3GPP) network and using the serving gateway to connect to other networks, EPS realizes the optimization of network deployment and the flexibility of capacity expansion. In addition to mobile network access, the 4G core network also supports video conferencing, 3D

television, and other applications that require high transmission speed. UEs in 4G networks can affort a transmission rate of up to 1 *Gbps*.

- 5G: service-based architecture (SBA)
In order to meet the needs of the three major application scenarios, 5G core network uses SBA [17] to split original network elements with multiple functions into multiple elements with independent functions, and each subnetwork can implement their own services independently. Compared with 4G networks, 5G core network can be reconfigured by software defined networks (SDNs) to realize virtualization and modularization of network functions. In addition, 5G core network supports network slicing, which can optimize business processes and data routing while enhancing network reliability. The expected theoretical speed of 5G network is as high as 20 *Gbps*. The high transmission speed can provide traffic support for various applications such as Virtual Reality (VR) / augmented reality (AR), Internet of vehicles (IoVs), UAVs, and smart city.

Through the discussion on the evolution of the core network, it can be seen that with the development of communication technology, the core network has the tendency toward network element function virtualization and network function modularization. With most of the hardware being generated in the virtualization domain, the network structure is easy to expand, shrink, and update. Therefore, the management problems caused by the excessive number of network elements can be greatly alleviated.

1.1.3 BACKHAULING

Backhauling [18] refers to the data transmission from BSs to the core network through a direct link or the Internet. The flexible and effective deployment of small cells is the premise of forming a network with multi-layer cells. Specifically, designing an appropriate backhauling scheme becomes the bottleneck for the successful deployment of UDNs. The backhauling of UDN can be divided into wireless backhauling and wired backhauling. Wireless backhauling has the advantage of flexible deployment, whereas wired backhauling has higher transmission reliability.

Wired Backhauling

- Copper wire
Copper wire resources have been deployed in the early stage of 2G and 3G construction. Due to the low cost, copper cables [19] were widely used. However, copper cables are subject to electromagnetic interference and serial interference, which affects the reliable transmission of data seriously. The backhauling solution with T1/E1 copper lines has the advantage of low latency and low jitter when dealing with voice and short messaging services, which can meet the requirements of synchronization in the system. Despite the convenience and low price, the expense of copper cables

increases linearly with the system capacity, making it difficult to adapt to the evolving communication systems.

- x digital subscriber line (xDSL)
  With the emergence of xDSL technology, copper wire has evolved from low-speed transmission of voice signals to high-speed transmission of data signals. In short, xDSL [20] technology is a broadband access technology based on the ordinary telephone lines. Moreover, xDSL technology can bind multiple DSL lines through multi-link bundling technology to meet the bandwidth requirements of backhauling networks. Although xDSL technology can reuse a great number of existing copper resources to reduce deployment costs, it is only suitable for networks with low traffic load.
- Optical fiber
  Optical fiber-based backhauling [21] is an ideal solution for wired backhauling with low unit bit transmission cost. In general, optical fiber has the advantages of low serial interference, anti-electromagnetic interference, high transmission quality, small size, and light weight. However, optical fiber is brittle in texture and has high requirements in power supplement.

Wireless Backhauling

Microwave wireless transmission is an effective alternative to wired backhaul, especially in complex-geographically areas. The frequency band of the microwave is $300\ MHz$–$300\ GHz$. In such a frequency range, unallocated frequency bands can be used to reduce cost in microwave transmission, while high frequency bands can be used to expand the capacity of the network. However, there are still some problems in microwave backhauling, especially in terms of short-distance transmission and multipath effects. In order to improve coverage, point-to-point transmission, point-to-multipoint transmission, and multi-hop configuration schemes are proposed for microwave backhauling.

- WiFi
  WiFi [22] is suitable for dense urban environments, as it has advantages not only in cost-efficiency, but also in providing high-quality services for multiple devices. WiFi can solve the problems of backhauling for small-scale cellular network traffic in authorized wireless bands. New WiFi technology integrates adaptive directional antenna with intelligent Mesh technology and channel prediction management technology.
- mmWave
  In order to further improve the wireless transmission rate, mmWave spectrum band has been considered for wireless backhauling [23]. Since the spectrum for mmWave is located within $60-100\ GHz$, mmWave has advantages of wide bandwidth and narrow beam. Besides, the channel quality of mmWave transmission is not effected by the climate conditions, and the system for mmWave transmission can be miniaturized due to small component size. Although mmWave is not suitable for long-distance backhauling,

it can be applied in the last-mile range backhauling networks to provide high-speed transmission rate.

Due to high bandwidth and stable transmission, optical fiber is the best solution for the backhauling. However, apart from requiring huge initial investment, the infrastructure construction of optical fiber backhauling is complex and time consuming. Although wireless backhauling is vulnerable to the propagation environment, it is flexible and cost-effective. Therefore, integrating wireless backhauling with optical fibers becoming a tendency to provide agile connection while ensuring stable and high-speed backhauling transmission.

1.2 ULTRA-DENSE NETWORK

1.2.1 DEFINITIONS OF UDN

With the development of wireless mobile networks, a variety of smart terminals and emerging applications have brought severe challenges to the system. On the one hand, the types of terminals that access to the network have expanded from traditional devices (e.g., mobile phones and computers) to ever-more-capable devices (e.g., smart wearable devices, autonomous vehicles, and UAVs), resulting in explosive growth of data traffic in communication systems. On the other hand, diverse applications including high-definition televisions, online video, and VR/AR have placed high requirements on the network bandwidth and the channel capacity in communication systems. 5G system is expected to support communication for more than 100 billion devices, with data transmission rate higher than 1 *Gbps* and transmission delay lower than 10 ms. The data proliferation urges the densification of base stations from macrocells, microcells, picocells, femtocells, to even tiny-cells.

Driven by the ever-increasing amount of mobile data, cellular networks evolve from small cell network to ultra-dense networks (UDNs), to provide high system capacity and spectrum efficiency. By bringing base stations (BSs) to the approximate spatial scale and number magnitude, UDNs would definitely bring unprecedented paradigm changes to the network design.

Up to now, there are two generally accepted definitions of UDN [24].

- *Definition 1: UDN refers to the cellular network where there are more cells than active UEs.*
- *Definition 2: UDN refers to the cellular network where the cell density is greater than 1000 cells/km^2.*

The small cells in UDN include femtocells, picocells, relays, and remote radio heads (RRHs), according to capacity, coverage, transmission power, and deployment scenarios. In particular, picocell is the small BS deployed in hotspot area (coverage area is within 100 *m*) and the load balancing can be achieved by offloading traffic on macro BS. Femtocell is the small BS deployed indoors to provide high-quality services for home UEs, whose coverage is less than 10 *m*. Relay is the access point deployed by the operator to cover the blind area, improving the cell-edge performance of macrocells. RRH is the RF unit deployed to extend the coverage of the

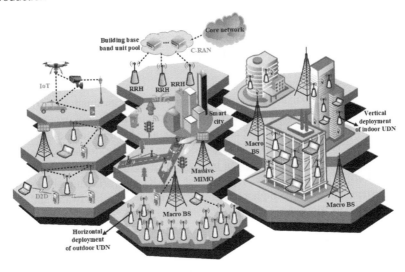

Figure 1.3   An example of a UDN includes V2V, D2D, IoT, and MIMO scenarios

central BS to a remote geographic location; different from picocell and femtocell, RRH can be used in centralized densification schemes [25].

As shown in Fig. 1.3, in smart cities, densification of cells can be realized in both the horizontal direction densification and the vertical direction densification. The horizontal direction densification means densely deploying cells on the horizontal plane of outdoor hotspots such as roads and building walls, while the vertical direction densification means densely deploying cells in houses, laboratories, offices or other indoor high-rise buildings. Figure 1.3 shows several representative UDN scenes, including cloud radio access network(C-RAN), device-to-device (D2D), IoT, and massive MIMO.

- C-RAN
  C-RAN originates from centralized cloud computing applications in the wireless networks [26]. The densification of C-RAN can be achieved by deploying multiple RRHs in the network. Compared with traditional cellular networks, dense C-RAN can significantly improve the spectrum efficiency and the energy efficiency of the system.
- D2D network
  D2D communication enables two neighboring devices to communicate directly without through a BS or a core network [27]. To further improve resource utilization, the spectrum partial reuse has been applied as an emerging technology in ultra-dense D2D network.
- IoT
  IoT interconnects computing devices, mechanical machines, objects, or people together. Massive access devices in the system can provide UEs with

high-quality services without manual interventions. Relying on AI, cloud computing, and other technologies, IoT can enable clumsy things with human intelligence [28]. The wide application of IoT would pose challenges on the deployment of UDNs.

- Massive-MIMO network
Massive-MIMO usually refers to the deployment of an antenna matrix consisting of a significant number of single antennas in BS. For example, BSs supporting massive-MIMO are usually equipped with hundreds of antennas to serve multiple UEs simultaneously in the same frequency resource. Therefore, Massive-MIMO can be regarded as a method of spatial densification [29]. Different from traditional MIMO networks, massive-MIMO system can achieve higher reliability with low-cost hardware and low-complexity algorithms.

1.2.2 CHARACTERISTICS OF UDNS

Compared with traditional networks, UDN has following different characteristics:

A. Dense and hierarchical cell deployment

Compared with traditional networks, the most intuitive characteristic of UDN is the extremely dense deployment of small cells and abundant types of small cells. In general, small cells in UDN include picocells, femtocells, relays, and RRHs. These cells can be divided into two types according to their functions: (1) Full-function cell: In a picocell or a femtocell, there is usually a full-featured BS, which can perform all functions of the macrocell; (2) Partial-function cell: As relay and RRH are extensions of the macrocell, they can extend the signal coverage effectively and perform all or part of the physical layer functions [30].

B. Super-frequent and complex handover

As compared with traditional multi-layer networks, mobility management in UDN is more complicated [31]. On the one hand, densely deployed cells have great differences in terms of transmission power, coverage, and network interface types. On the other hand, UE terminals are often in the coverage of multiple cells. When a user moves across the small and irregular cell coverage of BSs in UDNs, super-frequent and complex handovers are required among different network tiers, resulting in increased signaling overhead.

C. Challenging resource allocation

- Unpredictable channel state
Although it is difficult to obtain ideal channel state information (CSI) in a large network such as UDN, the known status information of multiple cells or all cells is a prerequisite for establishing an optimal resource allocation scheme. Considering that the acquisition of CSI requires huge signaling overhead, distributed resource allocation schemes can be applied to solve the problem. However, related schemes are only applicable to small-scale communication areas [32].

- Diverse QoE requirements
  With the emergence of new applications, QoE requirements of UEs are becoming increasingly diverse. Indicators such as latency, reliability, and transmission rate need to be considered in a customized and task-tailored manner in the resource management scheme.

D. Serious inter-cell interference

Although the large-scale deployment of small cells improves the performance of systems, at the same time, signals are subject to severe interference due to the closer distance with interfering sources during transmission. Inter-cell interference can be divided into two parts, which are the same-layer interference and cross-layer interference. The same-layer interference is caused by frequency sharing among cells from the same layer of base stations (e.g., the macro base stations) while the cross-layer interference occurs when cells from multiple layers of base stations (e.g., femto base stations and macro base stations) sharing frequency [33]. Interference coordination schemes can effectively mitigate interference and the compromise problems between interference and resource still require extensive research. Interference mitigation schemes need to be carefully designed to realize reliable and high-throughput transmissions.

E. Cell function expansion

With the development of information technology, various emerging applications appear in networks including virtual reality, augmented reality, intelligent transportation, and online games. In order to provide coherent and high-quality services, apart from basic communication functions, cells also need to integrate functions such as caching and computing.

- Caching
  Caching refers to storing copies of files in a cache, or temporary storage location, so that they can be accessed more quickly. Through storing files in BSs or UEs equipment dynamically, the resource load of wireless and backhaul during peak traffic hours can be further reduced. According to the proportion of storage files, caching can be divided into overall caching and partial caching. Implementing a caching scheme in UDN requires accurate prediction on content popularity. [34].
- Computing
  With the evolution of network systems, mobile computing has realized a transition from centralized cloud computing to mobile edge computing (MEC). MEC is a distributed computing framework enabling resource limited devices to offload their computational tasks to edge servers deployed on BSs or access points in cellular networks. MEC helps reduce network congestion and decrease latency, enhancing the quality of experience (QoE) for end users in networks, thereby is considered as an ideal enabling technology for Beyond 5G and 6G networks. [35].

F. Backhauling challenge

In UDN, it is difficult to establish ideal high-speed and low-latency backhauling for each small cell. Since backhauling has a great influence on cell capacity, it becomes a bottleneck for the successful deployment of dense networks. Compared with high-throughput and low-latency wired backhaul, wireless backhaul is a feasible solution in the complex network environment [36].

1.2.3 PERFORMANCE METRICS

* Coverage probability
  Coverage probability is the most common and important indicator in the performance analysis of wireless communication networks, which has the following definition [37].
  Definition: Coverage probability indicates the probability that a randomly selected UE can achieve the threshold of its signal to interference plus noise ratio (SINR).
  Coverage probability is also called success probability. Outage probability refers to the probability that the SINR of any UE goes below a specified threshold. If the SINR at the desired receiver is below a threshold, an outage occurs. Therefore, coverage probability, outage probability, and SINR distribution all can be used to quantify the link quality between the UE and the serving BS [38].

* Spectral efficiency
  Spectral efficiency refers to the average number of bits transmitted per second per unit of bandwidth. Cell spectral efficiency is used to measure the performance of a single cell. As the density of BSs increases, area spectral efficiency (ASE) is proposed as an important indicator for quantifying UDN performance. ASE is defined as the maximum data rate per unit bandwidth of random UEs using the same spectrum in the cell coverage area. The area in the ASE definition is related to the co-channel interference requirements of the cellular system, which independent of the target radio transmission characteristics [39].

* Energy efficiency
  Energy efficiency is a key performance indicator, which is defined as the ratio of network throughput or ASE to power consumption per unit area. The power consumption of BSs can be divided into two types, i.e., communication power consumption and equipment energy consumption. Specifically, communication power consumption of the BS is caused by sending and receiving signals within its coverage while equipment energy consumption refers to the energy consumed by the components of the BS, e.g., the power amplifier, the air conditioning, the rectifier, etc. Even if the transmission power of each small cell is low, the huge number of cells will consume a lot of energy [40].

- Fairness and priority

  Fairness is a key performance indicator for evaluating cell association, scheduling or resource management. The fairness of a scheme can ensure that each UE shares resources fairly, without the consideration about the average performance of the overall system. Max-min fairness, proportional fairness, weighted proportional fairness, and adaptive proportional fairness are common fair rules used in resource management. Conversely, priority ensures that UEs utilize individual resources based on their priority. Although the establishment of priorities makes resource allocation seem unfair, it can be seen as an extension of weighted fairness. With the popularity of UE-centric concept and the development of custom networks, the status of priority in performance indicators will become increasingly prominent [41].

1.3 OUTLINE

This book investigates the service and QoE provisioning in ultra-dense heterogeneous small cell deployment. In particular, in Chapter 1 we introduce UDNs by carefully defining UDNs regarding spatial deployment, generic characteristics of UDNs, and requirements of UDNs in order to ensure QoE of mobile users. In Chapter 2 & 3, we depict the resource management in UDNs among small cells in close proximity, mobility management for mobile users (address the super-frequent handovers), and interference management (dealing with the interference due to frequency-reuse in the vicinity). In Chapter 4, we study the enabling factors of UDNs, and the integration of UDNs with enabling technologies, such as massive-MIMO, cloud-RAN, mmWaves, D2D, and IoT. The promising applications driven by UDNs are illustrated in Chapter 5. Finally, we conclude the book and indicate future directions and challenges.

2 Resource and Interference Management

2.1 MODELING TECHNIQUES

2.1.1 STOCHASTIC GEOMETRY

Wireless networks are fundamentally limited by the intensity of the received signals and their interference [42]. Since both of these quantities depend on the spatial location of the nodes, it is necessary to establish a model that can truly reflect the randomness and spatial correlation of the BSs' distribution for the assessment of interference and critical performance. The diversification and randomization of UDNs require finding a rigorous, efficient, and easy way to analyze a point space distribution model to abstract the network.

In mathematics, stochastic geometry is the study of random spatial patterns [43]. It stems from the combination of integral geometry and point process theory, focusing on the study of random point patterns, such as the spatial point process (PP). Diverse point patterns provide the main building blocks for generating random object processes, allowing the construction of fine random space patterns. It is no exaggeration to say that stochastic geometry is one of the inherent statistical tools for modeling, analyzing, and designing UDNs with random node distribution.

Point Process (PP)

A PP describes the random process of the distribution of a set of random points located in n-dimension space, which is a mathematical model of phenomena or objects that can be represented as points in some types of space [44]. There are four main PP models used in the literature for modeling the wireless networks: Poisson point process (PPP), binomial point process (BPP), hard core point process (HCPP), and Poisson cluster process (PCP). The most basic one is the PPP and other stochastic processes can be regarded as the correction and optimization of the PPP.

A. PPP

The PPP is the most well-studied PP, and its importance stems from its ease of analysis. A stationary PPP with density λ is characterized by the following two properties:

- The number of points in any set B is a Poisson random variable with mean $\lambda |B|$. B here is an area, and $\mathbb{R}^d$ refers to the d-dimension space of interest. $\mathbb{R}^2$ is assumed to be a two-dimensional space
- The number of points in each disjoint set is an independent random variable

DOI: 10.1201/9781003148654-2

The probability of the number of points $N(B)$ located in the area B equal to k is given by:

$$Pr[N(B) = k] = exp(-\lambda |B|)\frac{(\lambda |B|)^k}{k!} \qquad (2.1)$$

The appealing feature of PPP is its invariance to a large number of key operations:

- The superposition of two PPPs with densities λ_1 and λ_2 results in a PPP with density $\lambda_1 + \lambda_2$
- The thinning of a PPP (i.e., selecting a point of the process with probability p independently of the other points and discarding it with probability $(1 - p)$) results in two independent PPPs of densities $p\lambda$ and $(1 - p)\lambda$. For example, using ALOHA as the MAC protocol in a wireless network leads to a thinning of the node set, and when the underlying nodes form a PPP, the resulting transmitter and receiver locations also form a PPP
- The probability distribution of $N(B)$ depends on the set B only through its size
- Conditioned on the number of points of Φ in a compact set B, the set of points $\Phi \cap B$ form a BPP. For a point, there are two probabilities: located in the area B or out of the area B, and the probability distribution of $N(B)$ depends on the set B only through its size $|B|$. Therefore, the number of points located in the area B follows a binomial distribution

B. BPP

The BPP is a PP obtained by placing n points $\Phi \in \{x_1, ..., x_n\} \subset \mathbb{R}^d$ independently and uniformly in a closed and bounded set $B \subset \mathbb{R}^d$. In the area A ($A \subset B$), the probability that the number of nodes is k ($k < n$) can be written as

$$Pr[\Phi(A) = k] = \binom{n}{k} \left(\frac{|A|}{|B|}\right)^k \left(1 - \frac{|A|}{|B|}\right)^{n-k} \qquad (2.2)$$

and a random point ξ in A is

$$P(\xi \in A) = \frac{v_d(A)}{v_d(B)} \qquad (2.3)$$

C. PCP

The PCP provides a tool for modeling spatial point patterns or spatial clustering with aggregated features. In practice, it is often referred to as the hierarchical model and is composed of two separate processes, i.e., the parent process and the daughter process. A PCP is usually constructed in the following steps:

- Define a parent PPP $\Phi_p = (x_1, x_2, ...)$ with density λ_p
- Form the clusters $K_{x_i} = K_i + x_i$ for each $x_i \in \Phi_p$, and the complete process is $\Phi = \bigcup_{x \in \Phi_p} K_x$

- The daughter points of the representative cluster K_0 are scattered independently with an identical spatial distribution:

$$F_{cl}(A) = \int_A f_{cl}(x)dx \qquad (2.4)$$

where $f_{cl}(x)$ is the probability density function of the cluster and $A \subset \mathbb{R}^2$.

D. HCPP

The characteristic of HCPP is that the distance between any two points must be more than a certain value (r_e). HCPP with an exclusion area of radius r_e is constructed in three steps:

- Define a parent PPP Φ_h in $\mathbb{R}^2$ with intensity λ_{ppp}
- Apply a mark M uniformly distributed in $[0, 1]$ to Φ_h
- Apply dependent thinning on Φ_h such that a point in Φ_h is retained in Φ_{HCPP} if it has the lowest mark in the disc with radius r_e

Combining the Voronoi tessellation model with the PPP to analyze cellular networks is a common method in current research. It is assumed that a BS and a UE obey the PPP distribution in space, and then use the Voronoi tessellation model to partition the space, which can reasonably describe the topology of the entire cellular network. Figure 2.1 illustrates a Voronoi tessellation network for PPP implementation of dense small cells. The red points represent the mobile UEs, and the black triangles represent the BSs. Each mobile UE associates with the nearest BS, in which cell boundaries are shown and form a Voronoi tessellation. This figure shows a simplest cellular network model.

In reality, the modeling effect of the PPP is not ideal, because points in PPP are completely independent and randomly distributed, but in the actual network there is spatial correlation, that is, there is mutual exclusion or attraction between points. In an actual BS deployment, the location of points usually presents a more regular and dense feature than PPP. In contrast, the emerging point processes such as BPP, HCPP, and PCP can more accurately simulate the location distribution of the network.

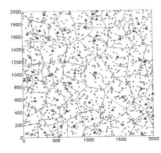

Figure 2.1   The Voronoi tessellation of a PPP realization of a cellular network with density $\lambda = 10 \ cells/km^2$ in an area of $2 \ km \times 2 \ km$

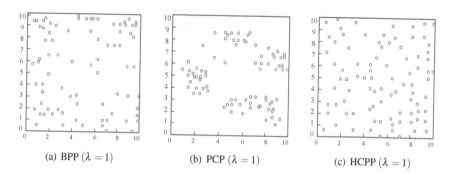

(a) BPP ($\lambda = 1$)   (b) PCP ($\lambda = 1$)   (c) HCPP ($\lambda = 1$)

Figure 2.2 Three different realizations of the PP

Figure 2.2 describes the BPP, PCP, and HCPP. The common feature of these three PP methods is that there is a certain correlation between points, which can capture the spatial characteristics of real network deployment. The defect of these three PP methods is that it is difficult to obtain theoretical analysis with guiding significance. The research of modeling technology focuses on the trade-off between modeling accuracy and analyzing ability, and finding a model that can characterize the key features of actual network node distribution.

Interference

Interference is anything that can modify or disrupt a signal as it travels along a channel between a source and a receiver. A device can receive information from a BS, denoted as downlink, and it can also receive information from other BSs, referred to interference. The interference measured at a point $y \in \mathbb{R}^d$ is

$$I(y) = \Sigma P_x h_x \ell(||y - x||) \tag{2.5}$$

where P_x is the transmit power of node x, h_x is the power fading coefficient, and ℓ is the loss function. When a transmitter generates a signal, it will deteriorate with the increasing path distance, and the path loss function is used to describe the path loss. If the signal power is assumed to follow Rayleigh fading, ℓ follows the exponential distribution. In some cases, $P_x h_x$ is regarded as a parameter F, which follows an exponential distribution. The important indicator closely related to interference, SINR, refers to the ratio of the strength of the received useful signal S to the strength of the received interference signal (noise σ^2 and interference I), which can be simply expressed as $SINR = \frac{S}{I + \sigma^2}$.

The primary result of stochastic geometry analysis is the average performance achieved by a network, the performance of randomly selected UEs or the average performance of all UEs, i.e., the average performance of UEs at all locations. This average performance indicator is expressed as spatial average performance. The spatial average performance metrics commonly used are listed as follows:

- Coverage/success probability and outage probability/SINR distribution
  This is a set of indicators that quantify the link quality between a UE and a serving BS, and is often used as a criterion for BS selection. Please refer to Section 1.2.3 for details.
- Ergodic rate
  The achievable transmission rate of a link is called the ergodic rate or capacity. According to Shannon theorem, the unit ergodic rate of a link can be expressed as:

$$\mathbb{E}[log_2(1+SINR)] = -\int log_2(1+x)dPr_c(x) \qquad (2.6)$$

  where $Pr_c(x) = Pr[SINR > x]$.
- Rate coverage and rate outage
  In small cell networks, the rate coverage is defined as the probability that the achievable rate of an arbitrary UE above a certain minimum. Conversely, the rate outage is the probability that the achievable rate of an arbitrary UE falls below a certain threshold [45].
- Average spectral efficiency
  The average number of transmitted bits per second per unit bandwidth represents the efficiency of the spectrum [46]. Also, the cell spectral efficiency is another form of this metric to measure the performance of a single cell.
- Area spectral efficiency
  Densification of cellular networks increases the reuse of spectrum per unit area. Thus, area spectral efficiency is an important metric to quantify the performance of UDNs. ASE is defined as the average achievable data rate per unit bandwidth per unit area [47].
- Energy efficiency
  It is a performance indicator that measures the benefit-cost ratio by comparing achievable rates with energy costs [48]. Please refer to Section 1.2.3 for details.

Laplace Transform

The Laplace transform is a widely used integral transform in mathematics. It can be used to transform ordinary differential equations into algebraic equations, which makes it easier to solve ordinary differential equations [49]. In engineering, the importance of the Laplace transform is to transform a function of time into a function of complex frequency. It is defined by

$$F(s) = \int_0^\infty e^{-st} f(t)dt \qquad (2.7)$$

where $t \geq 0$ and $s = \sigma + i\omega$, with real numbers σ and ω. For example, the Laplace transform for $f(t) = e^k t$ can be obtained by the following procedure:

$$\mathscr{L}[f(t)] = \int_0^\infty e^{kt} e^{-st} dt = \int_0^\infty e^{-(s-k)t} = \frac{1}{s-k}$$

An Application of Stochastic Geometry in Handover Management

The reduction in the distance between BSs in UDN triggers frequent handovers while increasing spectrum utilization and network capacity. The delay caused by handover is likely to reduce or even offset the benefits of network densification. In order to reduce frequent handovers, a handover scheme called handover skipping is proposed in which a UE does not always maintain an optimal connection, but instead chooses between connecting to its best serving BS and skipping the following one alternately [50]. A mathematical paradigm based on stochastic geometry is developed to quantify the effect of handover delay on the average ergodic rate in a UDN.

This study primarily considers a single handover skipping scheme in a single-tier cellular network, which consists of BSs arranged according to the PPP. In the best-connected case, each mobile UE is associated with the nearest BS, i.e., a UE is connected to the BS in the Voronoi cell where it is located. In the blackout case, a mobile UE skips the nearest BS, and the second and third nearest BSs are selected to serve it cooperatively. According to the correlation of the model, the schematic diagram of the skipping scheme can be obtained, as shown in Fig. 2.3.

The Laplace transform of I_r in the best connected case is given by:

$$\mathscr{L}_{I_r}(s) = \int_0^{r_0} \frac{\mu}{\mu + sr^\alpha} f_{r_1}(r) dr \tag{2.8}$$

where $f_{r_1}(r) = 2\lambda \pi r e^{-\lambda \pi r^2}$, which is the expression of the distance distribution between the best connected UE and its serving BS. For the blackout case :

$$\mathscr{L}_{I_r}(r) = \int_0^{r_0} \frac{\mu}{\mu + sr^\alpha} f_{r_2}^{(bk)}(r|r_1) dr \tag{2.9}$$

where $f_{r_2}^{(bk)}(r|r_1) = \frac{2r}{r_1^2}$ is the expression of the distance between the UE and the skipped BS in the blackout case. Analytical expressions and simulation results show that the single handover skipping scheme improves the long-term average rate of mobile UEs in UDN.

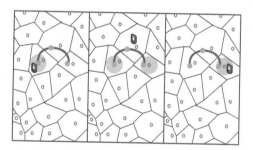

Figure 2.3 Schematic diagram of skipping scheme: the blue circle represents the BS following the PPP, the red arc indicates the UE's movement trajectory, and the green circle represents the selected serving BS

An Application of Stochastic Geometry in Energy Efficiency

Densification of cells inevitably increase energy consumption, which is an issue that should be considered during the BS deployment phase. Recently, Zhang *et al.* [51] discussed the impact of BS deployment on energy efficiency in UDNs using stochastic geometry.

The study mainly considers the downlink ultra dense HetNet. The spatial distribution of the BSs in the kth tier is modeled as a homogeneous PPP with a density of λ_k in the two-dimensional Euclidean plane. The coverage area of BSs forms a weighted Poisson Voronoi Tessellation due to the difference in transmit power of a BS across the layers. UEs follow an independent PPP with a density of λ_u. According to PPP, the average number of UEs served by the BS in the kth tier, the coverage probability of the kth tier, and the minimum achievable data rate of the labeled UE associated with the kth tier BS are obtained. Thereby the minimum achievable throughput of the entire UDNs can also be obtained. Its ratio to total power consumption is the energy efficiency, i.e.,

$$\eta_{EE} = \frac{R_{total}}{\sum_{k=1}^{K} \lambda_k P_k} \tag{2.10}$$

Expressions and simulations show that as the density of BS in UDNs increases, the energy efficiency will increase first and then decrease. Therefore, in UDNs, the density of the BS should be carefully designed. Otherwise, excessively dense deployment of small cell BSs will weaken network energy efficiency.

2.1.2 GAME THEORY

Wireless networks typically involve multiple UEs competing for resources, such as bandwidth and transmission power. Because of the selfish behavior, each UE wants to meet its communication requirement best, and this disorderly competition can lead to a decline in the overall network performance. At the same time, due to the time-varying characteristics of wireless channels, the stability and time consistency of decision-making should be considered [52]. The great number of nodes in UDNs undoubtedly intensifies the competition. Therefore, how to stimulate selfish node collaboration is a problem that needs to be solved.

Game theory is the study of mathematical models of strategic interaction between rational decision-makers. In particular, it focuses on decisions in environments where each player's decision can affect the outcome of other players. It predicts how a player behaves in a strategic environment, or as a behavioral plan for the player, telling the players how they should act [53]. In interactive scenarios such as resource management and interference avoidance, using game theory to study the cooperation and incentive methods of selfish nodes is an effective way to optimize network performance and maximize resource utilization.

Representation of Games

The expression of a game includes the strategic game and the extended game. The former is a kind of normative description of the game problem, i.e., the standard game. The latter is a decision model based on the assumption that each player chooses only one action or strategy and the players choose at the same time. The difference between the two is mainly in whether to describe the sequence of game actions.

The strategic game is a static model, which is generally suitable for describing the static game with complete information without considering the game process. The key elements are as follows:

- A set of players (at least two players): $N =$ {Player 1, Player 2, ... , Player n}
- Each player's strategies/actions set: $S_i, \forall i \in N$
- Payoffs received by each player for the combinations of the strategies, or for each player, preferences over the combinations of the strategies: $u_i(s_1, s_2, ...s_n)$, for all $s_1 \in S_1, s_2 \in S_2, ..., s_n \in S_n$

Compared with the strategic game that mainly describes the game results, the extended game focuses on the analysis of the sequence structure of the players when they encounter the decision problems in the process of the game. The elements are as follows:

- A set of players
- The order of the players' actions, i.e., when each player acts
- Sequence structure, the decision problems faced by each player in the course of action, including the action options available to the player, and the information learned
- The payoff function of the player

Nash Equilibrium

Nash equilibrium, named after *John Forbes Nash Jr.*, is used to analyze the outcome of the strategic interaction of multiple players in a game. In a Nash equilibrium, no player can increase its expected gain by changing its own strategy while the other players keep their strategies unchanged. Formally, in the game $\{S_1, S_2, ..., S_n, u_1, u_2, ..., u_n\}$, a strategy profile $(s_1^*, ..., s_n^*)$ is a Nash equilibrium if

$$u_i(s_1^*, ..., s_{i-1}^*, s_i^*, s_{i+1}^*, ..., s_n^*) \geq u_i(s_1^*, ..., s_{i-1}^*, s_i, s_{i+1}^*, ..., s_n^*) \qquad (2.11)$$

for all $s_i \in S_i$. That is, the strategy s_i^* of any player i is the best countermeasure for the combination of other players' strategies $(s_1^*, ..., s_{i-1}^*, s_{i+1}^*, ..., s_n^*)$. Note that, there can be more than one Nash equilibrium in a game.

Example 2.1.1 *An example is the* prisoner's dilemma, *which reflects that individual best choice is not group best choice. Two suspects held in separate cells are charged with a major crime. However, there is not enough evidence. Both suspects are told the following policy:*

- *If neither confesses, then both will be convicted of a minor offense and sentenced to one month in jail*

- *If both confess, then both will be sentenced to jail for six months*
- *If one confesses but the other does not, then the confessor will be released but the other will be sentenced to jail for nine months*

As a result, the two men at the same time in confessions or not confessions dilemma. But the two people could not communicate, so from the angle of their own interests, they both chose to confess according to their own rationality. The strategy of not being confessed and thus sentenced to one month, which is beneficial to both parties, has not been chosen. This is called the Nash equilibrium. In this case, no player can "act alone" (that is, unilaterally change decisions) and gain more.

Game Types

Players in the prisoners' dilemma have no means of communication. Thus they cannot agree on how to play the game. This kind of game is called the non-cooperative game, which is classified depending on the information exchange and the existence of binding executable contract. In addition, the game can also be divided based on the player's action order or the other players' information.

- The cooperative game / The non-cooperative game
  The cooperative game refers to the game with competition between coalitions. In each coalition, multiple players cooperate with each other due to the external enforcement or binding agreements. The cooperative game focuses on coalition forming and profit distribution. The non-cooperative game refers to the game where the players can not form coalitions and have to compete independently. The non-cooperative game focuses on individual strategy and payoff and Nash equilibrium analyzing.
  If the two players in the prisoner's dilemma sign an agreement before the game, both of them promise to choose not to confess. In order to make sure that the promise is fulfilled, both players pay a third party a margin worth more than the excess returns from a deviation from the strategy combination, and if anyone violates the agreement, the margin is waived. With such an agreement, everyone's earnings will be improved. Through a binding agreement, previously impossible without cooperation programs can be achieved.
- The static game / The dynamic game
  Depending on the time sequence of the player's behavior, the game can be divided into static games and dynamic games. In a static game, players act without knowing what other players choose to do before they act. In a dynamic game, players' actions are in sequential order, and later actors get the action information of the previous players.
  The prisoner's dilemma is a static game, and the repeated prisoner's dilemma is a dynamic game. By repeating, the player can decide the current choice according to the opponent's past choices, and each player has the opportunity to "punish" another player's uncooperative behavior in the previous round. The best strategy is considered to be a "tit-for-tat" strategy,

which is to cooperate at the beginning of the repeated game and then adopt the strategy of the opponent's previous round.

- The complete information game / The incomplete information game
  Depending on the degree of known information by players, the game can be divided into the complete information game and the incomplete information game. A complete information game means that each player has accurate information about the characteristics, strategies, and benefit functions of all other players. In the incomplete information game, the players are not fully aware of some information about the game.

The prisoner's dilemma is a complete information game, which can be transformed into an incomplete information game. It is assumed that prisoner A follows two types of behavior, i.e., rational behavior and irrational behavior, with probabilities of $1 - p$ and p respectively, and that prisoner B has only one type – rational behavior. It is assumed that rational prisoners can choose arbitrary strategies. Non-rational prisoners have only one "tit-for-tat" strategy. For a game, no matter whether the prisoner A is rational or not, the confession strategy is always the best for the prisoner B.

An Application of Game Theory in Resource Management

Resource management problem (such as power control and UE scheduling) becomes challenging due to the large scale of UDNs. Due to the participation of a large number of devices and the severe coupling between their control parameters, the traditional methods in small-scale network deployment are not sufficient to study resource optimization in UDNs. In addition, the resource management challenges in UDNs will be further aggravated by the fluctuation of space-time traffic requirements, the dynamics of channel conditions and the increased overhead due to the need for coordination. For example, the uncertainty of queue state information (QSI) and CSI and their evolution over time play a key role in resource optimization.

A novel approach for joint power control and UE scheduling is proposed in UNDs [54]. The energy efficiency is formulated as a dynamic stochastic game between small BSs due to severe coupling in interference. As shown in Fig. 2.4, as the number of interfering small BSs grows, the interference observed at a generic UE becomes independent of individual states and transmission policies of small BSs and only depends on the time t and the limiting distribution $\rho(t)$. In such a scenario, dynamic stochastic game can be solved by mean field game. Mean field game is the study of strategic decision making in very large populations of small interacting agents [55]. In continuous time, a mean field game is typically composed of a Hamilton-Jacobi-Bellman (HJB) equation and a Fokker-Planck-Kolmogorov (FPK) equation. Under fairly general assumptions, a mean field game is the limit as $N \to \infty$ of an N-player Nash equilibrium [56]. UE scheduling is formulated as a stochastic optimization problem and solved by using the drift plus penalty approach in the framework of Lyapunov optimization.

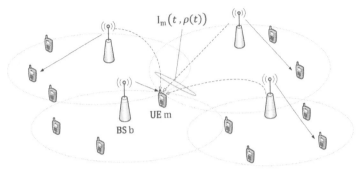

Figure 2.4 Mean Field Interference

For a given UE scheduling mechanism, the power control problem can be formulated as a dynamic stochastic game

$$\varsigma = (N, \{S_i\}_{i \in N}, \{X_i\}_{i \in N}, \{u_i\}_{i \in N}) \tag{2.12}$$

where N is the set of players, i.e., the small BSs, and S_i is the set of actions of player i, i.e., the choices of transmit power p_i for given scheduled UEs. In addition, X_i is the state space of player i, such as QSI q_i and CSI h_i, and u_i is the average utility of player i that depends on the state transition $x(t) \to x(T)$ as

$$u_i(t, x(t)) = \mathbb{E}\left[\int_t^T u f_i(\tau, x(\tau), Y(\tau)) d\tau\right] \tag{2.13}$$

where $u f_i$ is a utility function of player i, and $Y(t) = [y_i(t), y_{-i}(t)]$, $y_i(t)$ is the small BS local control vector, in which $y_{-i}(t)$ is the control vector of interfering small BS. To solve the original dynamic stochastic game with N small BSs, N HJB partial differential equations (PDEs) must be solved. The solution of the mean field game with a large N can be found by simply solving two coupled HJB-FPK PDEs. Mean field game reduces the complexity considerably and can scale well with the number of players. When considering multiple UEs associated with a small BS, UE scheduling directly affects the performance of each small BS. Due to the dynamic characteristics of the channel and any arrival, the UE scheduling process becomes a random optimization problem. In order to solve the stochastic optimization problem of each small BS i, the drift plus penalty method in the Lyapunov optimization framework can be applied to decompose the stochastic optimization problem into sub-strategies that can be implemented in a distributed manner. In this way, UE scheduling and power control are performed through game theory, and the transmission power of each node not only achieves smooth transmission, but also does not cause excessive interference to surrounding nodes.

2.2 RESOURCE MANAGEMENT IN ULTRA-DENSE HETEROGENEOUS NETWORK

2.2.1 QOE PROVISIONING TECHNIQUES WITH OFDMA OR NOMA

OFDMA

Orthogonal frequency division multiple access (OFDMA) is a multi-user version of OFDM digital modulation scheme [57]. OFDMA takes advantages of OFDM to assign channels to subcarriers, and then allocates different sets of orthogonal subcarriers to different UEs dynamically to achieve simultaneous low-data-rate transmission from several users [58].

As a modulation method, the principle of OFDM is to divide the total bandwidth of the whole system into several orthogonal sub-channels and convert high-speed data signals into parallel low-speed sub-data streams, which are modulated on mutually orthogonal subcarriers respectively to achieve synchronous transmission of signals in sub-channels. In traditional frequency division multiplexing (FDM) systems, the spectrum of multiple subbands does not overlap with each other, and there is usually an interband protection bandwidth between the two subbands to avoid interference [59]. However, in OFDM systems, subbands are orthogonal to each other and signals can be received without spectrum separation, which improves the utilization rate of the spectrum. Moreover, narrowband transmission is carried out on each sub-channel, then inter-carrier interference can be greatly eliminated. In practical systems, using Fast Fourier transform to deal with orthogonal subcarriers can significantly reduce the complexity of computation. OFDM can combine diversity, space-time coding, interference suppression, and smart antenna technologies to maximize system performance. On this basis, the multi-UEs OFDMA system can provide high spectrum efficiency through adaptive subcarrier selection and power allocation, which improves the performance of anti-interference ability and data transmission rate.

In practical systems, using Fast Fourier transform to deal with orthogonal subcarriers can significantly reduce the complexity of computation. OFDM can combine diversity, space-time coding, interference suppression and smart antenna technologies to maximize system performance, which has become one of the key technologies in the 4G evolution. On this basis, the multi-UEs OFDMA system can provide higher spectrum efficiency through adaptive subcarrier selection and power allocation, which improves the performance of anti-interference ability and data transmission rate.

NOMA

As the growing demand for high-speed data services and large-scale IoT device access, OFDMA technology can barely meet the requirements of future wireless communication systems. In order to guarantee the performance of communication systems in the EMB, mMTC and URLLC scenarios, academia and industry are gradually transiting their research direction from OFDMA to NOMA [60]. As shown in

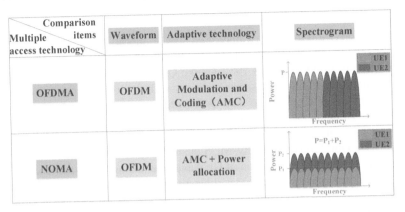

Figure 2.5   Some differences between NOMA and OFDMA

Fig. 2.5, NOMA has better development potential than OFDMA. Here we introduce two common NOMA technologies.

- Power-domain NOMA
  Power-domain NOMA technology rises from OFDMA [61]. OFDMA technology is applied between the non-orthogonally sub-channels in NOMA system, so that sub-channel transmission does not incur interference to each other. There are two key technologies in power-domain NOMA, i.e., power allocation technology and successive interference cancellation technology [62].
- Code-domain NOMA
  In the code-domain NOMA system, multiple UEs can share the same resources at the same time, so that the communication system is not strictly limited by the number of available resources [63]. In order to avoid resource allocation conflict in the process of transmission, an extension code is used to transfer resources. At present, the code domain-based NOMA technology mainly has the following types: multi-UE shared access (MUSA) [64], sparse code multiple access (SCMA) [65], and pattern division multiple access (PDMA) [66].
  SCMA, proposed by Huawei, is an enhanced low-density spread spectrum technology. By using low-density or sparse non-zero-element sequences, the complexity of message passing algorithm at the receiver can be greatly reduced [67]. Compared with OFDMA, the SCMA system can serve multiple UEs at the same time, and further improves the system capacity. Moreover, the SCMA system can flexibly design the codebook according to the diversity of future communication scenarios to meet different UE requirements.

QoE

Quality of service (QoS) is a widely used service metric, to measure the network performance such as network packet loss rate, throughput, delay, and bit error rate from a technical point of view [68]. Unlike QoS, Quality of QoE also comprises non-technical aspects, which is more relevant with the users' perception. To facilitate analysis, it is necessary to map QoE with numerical values to quantify QoE. One of the common methods is mean opinion score (MOS), which maps QoE to a function with multiple QoS parameters and uses different MOS values to characterize different QoE levels. Generally, MOS quantifies QoE from poor to excellent into five levels, making the description of UE subjective feelings more convincing.

QoE evaluation methods includes subjective evaluation and objective evaluation. The subjective evaluation method refers to a group of UEs directly performing QoE evaluation on service in a certain environment. The subjective evaluation is a direct and accurate evaluation method, while it cannot be widely used due to its has the issues of strict test environment requirements, poor portability, and high cost. The objective evaluation method does not require UEs to participate in an evaluation process, but establishes a QoE model based on the key QoS parameters. For example, when a voice service is evaluated, a unique evaluation model of QoE can be established through parameters such as delay and throughput. Compared with the subjective evaluation method, the objective evaluation method is appropriate for mathematical modeling.

An Application of QoE Provisioning Technique with NOMA

In UDN, the dense deployment of small BSs poses challenges on the system spectrum resources. A resource allocation strategy for multi-cell NOMA networks [69] is proposed to improve the QoE of UEs in the communication system. Figure 2.6 shows a multi-cell downlink NOMA transmission scenario, where the transmit and receive antennas are equipped on both UEs and BSs. The whole bandwidth is available on each cell, which is divided into multiple equal sub-channels. A sub-channel is shared by multiple UEs associated with the same BS, and it is assumed that BSs work cooperatively to serve the UEs whose CSI [70] are available to the BSs. Moreover, the use of multiple-BS diversity in terms of power allocation and UEs scheduling enhances the reliability of UE data reception.

The interference received by a UE is mainly from other UEs in the same sub-channel, and SIC can be carried out at the UEs with stronger equivalent channel gains in the NOMA system. Taking the web browsing application as an example, the MOS function obtained by modeling QoE is proportional to the data transmission rate of UE on each sub-channel at different BSs, and is related to the size of the web page.

The goal of the scheme is to maximize the sum MOS of UEs in the network, but UE-BS association, sub-channel assignment, and the power allocation are coupled to each other in terms of the UE data rate. In order to get the solution effectively, the problem is decomposed into two subproblems, one is the problem of UE-BS association and sub-channel allocation, the other is power allocation between UEs.

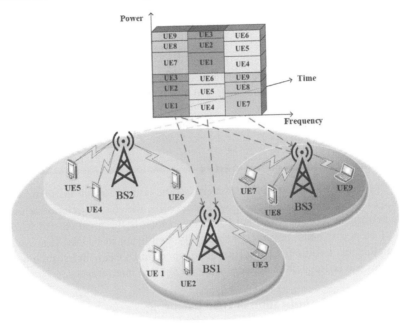

Figure 2.6 A multi-cell downlink NOMA transmission scenario

- UE-BS association and sub-channel assignment

  The combinational optimization problem among UEs, BSs, and sub-channels is NP-hard. To find the optimal solution, it is decomposed into two two-dimensional matching subproblems which are UE-BS matching problems and (UE, BS)-SC matching problems. According to the speed of achievable date rate, UEs and BSs establish preference lists among BSs and UE sets respectively, then the extended deferred acceptance based algorithm is utilized to construct an initial matching state between UEs and BSs. Based on the equal allocation of UE's transmission power, after the construction of swap operation, the final matching lists of UEs and BSs can be obtained. Similar to the method mentioned above, preference lists of BSs and sub-channels can be established based on sub-channels and BSs sets, respectively. After getting the matching result of sub-channels and (UE, BS) units, the solution of 3D matching about (UE, BS, SC) units can be obtained.

- Power allocation optimization

  As the UE-BS association and the SC assignment are known, the sum MOS maximization problem can be simplified to a power allocation optimization problem, the restrictions of which are equivalent to channel gains constraint and sum power constraint. As the objective function is non-convex, a suboptimal approach based on the successive convex approximation [71] technique is proposed to convert it into a convex optimization problem, which can be effectively solved by standard convex solvers.

By comparing the performance of the proposed scheme in the OMA [72] system and the NOMA system, it can be seen that the NOMA technique achieves higher QoE level. In general, the NOMA system has high spectral efficiency, and can guarantee fairness between UEs.

2.2.2 LOAD BALANCING

Motivation for Load Balancing in UDN

In UDNs, the densification of BSs exacerbates the issue of traffic fluctuation in the coverage of small cells as compared with large cells, which may deteriorate the network performance [73]. Thus load balancing, which is the process of distributing traffic load across cells, becomes increasingly significant in UDNs. However, it is difficult to apply traditional load balancing methods in UDNs with irregular network topologies, super-frequent handovers, and complicated interferences.

Definition and Quantification of Load Balancing

Load balancing is a key technology for resource management in a heterogeneous wireless network. A typical scenario of load balancing is shown in Fig. 2.7. By coordinating radio resources between different cells, efficient utilization of

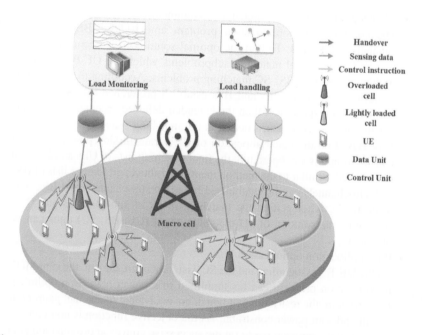

Figure 2.7 Application of load balancing technique in UDN scenario

communication resources can be achieved. Specifically, the network console performs load balancing operation on cells according to the load information of different cells to reduce the number of overloaded cells. In order to describe the use of resources in cell, the load level index can be used to characterize the load status. The cell load level index is defined as L_i:

$$L_i = w_1 U_i + w_2 R_i + w_3 T_i \qquad (2.14)$$

where U_i is the ratio of UEs' number in cell i to the maximum UE capacity of the cell, R_i is the ratio of physical resource blocks number used by the cell i to the total number of physical resource blocks, and T_i is the ratio of the real-time traffic in cell i to the total traffic. w_1, w_2, and w_3 are the weights of U_i, R_i, and T_i, respectively. Considering that a large number of cells are distributed in the entire network system, the average load L can be obtained by L_i. L_i can still measure the load status of cells in multi-layer heterogeneous network system, however, different load thresholds is set for different types of cells to determine whether a cell is overloaded.

Load Balancing Based Scheme

The strategy of load balancing can be divided into channel borrowing [74] and traffic offloading [75].

- Channel borrowing
  Channel borrowing enables heavily loaded cells to borrow unused channels from lightly loaded cells, or occupy the shared channels with high priority.
- Traffic offloading
  Traffic offloading refers to the strategy where some UEs in heavily loaded cells can be directly transferred to lightly loaded cells to improve the overall QoE.

In heterogeneous UDNs, a UE is likely to receive higher transmission power from amacro BS than from a small BS. On the other hand, heavy load on the macro BS affects system capacity and surely lower the data rate on UEs. Therefore, load balancing methods in heterogeneous UDNs need to comprehensively consider the current data rate, the future handover rate, and the complicated interferences. Currently, the methods in load balancing of UDN are mainly based on combinatorial optimization, game theory, Markov decision process (MDP), etc.

- MDP
  In UDNs, the handover scheme is complicated and a long-term decision-making process due to the coexistence of multiple radio access techniques in UDNs and irregular coverage area of BSs. MDP provides a solution for the continuous optimization of discrete-time stochastic systems with uncertainties. The goal of MDP is to maximize future expected rewards after performing operations in the current state. According to the known load conditions of BSs, MDP can help UEs make advisable handover decisions. Although it is difficult to define models for state and state transition

in large-scale heterogeneous systems, MDP can provide a rational solution for self-organizing networks with the combination of centralized design and distributed design through the control of decision makers [76].

- Game theory

  In UDNs, UEs compete the limited system capacity with each other in order to obtain high transmission rate. Game theory allow systems to analyze interactive decision-making processes and provide manageable solutions for complex decentralized optimization problems. Models such as Cournot model can therefore be used to explore the load balancing problem in UDNs. Through Nash equilibrium in the game model, the theoretical optimal load balancing solution can be obtained. However, the optimality of the solution cannot always be guaranteed [77].

- Cell range expansion

  Cell range expansion (CRE) is a biasing association method for load balancing proposed by 3GPP. When cell selection is performed based on maximum SINR or reference signal received power (RSRP), adding a positive bias to the SINR or RSRP received by small BSs can partially offset the power advantage of macro BSs, resulting in a fair resource allocation for small cells. That is, the coverage of small BSs gets virtually expanded. The offload of traffic from macro BSs to low-power small BSs can effectively improve the resource utilization and alleviate the uplink interference correspondingly [78].

An Application of Load Balancing

A simple and effective technique for traffic offloading from macro cell to small cell is biasing association. A low complexity algorithm for cell-specific bias calculation in UDN is proposed [79]. Figure 2.8 shows a downlink transmission in heterogeneous wireless networks with a macro cell and several small cells. The distribution of small cells and UEs are consistent with independent uniform spatial PPP. In the case of co-channel deployment, both macro cell and small cells use the same frequency band, while in the case of orthogonal deployment, macro cell uses different frequency band compared with small cells and generate no interference on small cells tier.

The communication resources in terms of physical resource blocks can be assigned to users with high QoS requirements. To serve more UEs, each cell would choose a bias value to expand the advantage in SINR-based connection selection. However, the physical resource blocks of each cell are limited. If the cell cannot provide enough physical resource blocks to its served UEs, the connection will be interrupted. Therefore, appropriate bias values should be selected for each cell to minimize the total outage-rate of the system.

Using the method of Gibbs sampling can solve the NP-hard problem of finding cell-specific optimum bias values [80]. In order to reduce the management complexity, a suboptimal algorithm is proposed, where a lightly loaded cell increases its bias to get partial traffic from its neighbor overloaded cell. The algorithm has two steps:

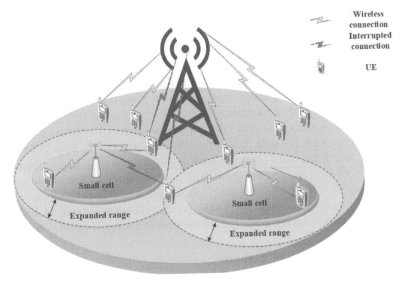

Figure 2.8 Load balancing in a downlink heterogeneous network transmission scenario

Step 1: Initial bias calculation

- From the reference signals broadcasted by neighboring cells, UEs measure the average SINR which is then reported to potential serving cells
- Assuming all cells are initially unbiased, each cell independently calculates the maximum bias that can be set to avoid overload, and then sets it as initial bias

Step 2: Bias refinement

- Each cell sends its initial bias value to the neighbor cell
- Considering the initial bias value of the neighbor cell, each cell calculates the maximum bias once more
- Cells broadcast the final bias value to the UEs
- UEs connect to the cell with the highest bias received SINR

Compared with Gibbs sampling, the proposed algorithm can achieve near-optimum solution with lower complexity and less signalling cost as compared with the benchmarks. Although CRE is always achieved by UEs association, the motivation behind CRE is improving the network performance through load balancing. Among multiple load balancing techniques, CRE has advantages of low system cost and its adaptation to heterogeneous networks.

2.2.3 ENERGY-SAVING TECHNIQUES

Motivation for Energy Saving in UDN

The densification of small cells and the resultant complex transmission technologies would bring large amount of energy consumption. Effective techniques need to be designed to reduce the energy consumption by controlling the power saving mode of BSs considering the tidal effect of data traffic.

Moreover, considering the tidal effect in the communication network, when the traffic load is reduced, the state that all the BSs are still working leads to an excess of capacity, which increases energy consumption and cost. Nowadays, the energy consumption of BSs account for about 80% of the entire communication system. With the densification of BSs, the energy consumption of BSs is rapidly increasing at an annual growth rate of 30%−40% [81]. As a result, energy saving for BSs is the main target of energy saving techniques in the current communication system.

Energy Consumption in BS System

In particular, the energy consumption of a BS consists of the following three components:

- The communications equipment consumes 49%−51% of the total energy consumption
- The heating and cooling system of a BS consumes 40%−46% of the total energy consumption
- Power and auxiliary equipment, which include monitoring equipment and lighting equipment, consume 6%−8% of the total energy consumption

Specifically, the energy consumption of the BS can be divided into two parts: static energy consumption and dynamic energy consumption [82].

- Static energy consumption
  The static energy consumption occurs when the BS is in an idle state and does not provide services to any active UEs, which mainly includes the power supplementary consumption of the main equipment and the running consumption of the environmental equipment.
- Dynamic energy consumption
  The dynamic energy consumption includes the power consumption of the power amplifier, feedback loss, etc. Dynamic energy consumption usually increases with the rise of the traffic load.

Energy-Efficient Scheme

There are three common energy saving techniques for BSs: (1) Using green renewable energy, such as wind, tidal, solar, and geothermal energy; (2) Optimizing the network architecture and adjusting the deployment of BSs in the network; (3) Controlling the power saving mode of BSs according to the traffic requests.

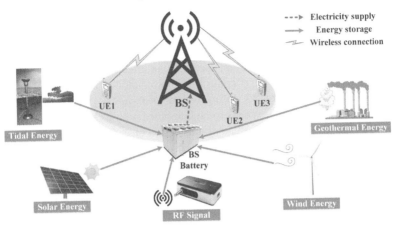

Figure 2.9 The application of energy harvesting technology in UDN

- Energy harvesting

  Wireless devices on BSs are usually battery-powered and consume much energy during data transmission and processing. As shown in Fig. 2.9, a feasible solution is energy harvesting technique, which enables devices to harvest renewable energy from the surrounding environment to power the communications of BSs [83]. In a solar-energy powered system, the generated power is fed back to the charge regulator, which is responsible for powering the BS and directs excess power to energy storage system. As long as the power generated by the renewable energy source is not sufficient to operate the equipment, BSs can directly use the energy stored in the battery [84]. However, relying on external energy sources increases the cost of the network. Hence an energy harvesting method based on RF signals has been proposed. In this case, wireless signals can be used to convey both information and power, while the functions cannot be implemented simultaneously in actual circuits [85]. To solve this problem, a receiver mechanism based on power allocation and time switching is proposed, which enables energy collection and information processing to perform at different power domains and time intervals.

- BS deployment

  In UDN, the dense deployment of small BSs would serve a large number of users, while resulting in additional energy consumption as well as radio interference [86]. The number and location of small BSs in the network need to be carefully planned. There are multiple BS deployment strategies, including Monte Carlo simulation [87], stochastic geometry theory, and heuristic algorithm [88]. The well-known simulated annealing using Monte Carlo simulations can assess the performance of networks under a given BS deployment. Stochastic geometry based methods can analyze the network performance in terms of energy efficiency, average rate, etc, considering various BS deployment, which can inversely optimize network parameters, e.g., BS intensity, to achieve optimum energy efficiency.

- Cell sleeping
Cell sleeping technique is considered as an effective way to improve network energy efficiency by controlling power mode of some components, devices, or the entire BS during the off-peak hours [89]. Generally, the cell sleeping technique needs to be combined with techniques such as cell zooming [90] and relay transmission [91]. For example, when a BS with low load is turned off, its neighbor BSs need to expand their coverage to achieve coverage supplementation. Moreover, if the expansion is insufficient, a relay transmission technique can guarantee high-quality communication requirements. Currently, cell sleeping technique includes centralized architecture and distributed architecture. In the centralized architecture, the system relies on the collection and calculation modules to obtain the optimal sleeping strategy. Although the global optimal solution scheme improves the system performance, UEs in this mode are passive and have high probability of outage. In the distributed architecture, the system does not need additional equipment to collect information and process them. UEs are in proactive states and can select the BS with the best signal to connect. Although the optimal scheme in this mode is locally optimum, the time complexity of the algorithm implementation is relatively low [92].

An Energy-Efficient Method

In order to meet the communication requirements of UEs, a caching technique is introduced at the edge of the network to offload the backhaul. Apart from improving transmission efficiency, a caching technique can reduce the energy consumption in the core network. In order to achieve energy efficiency, cell sleeping and caching technique can both be adopted in the network.

A UE connection matrix-based cell sleeping algorithm with edge caching is proposed [93] to reduce the energy consumption of the whole system. As shown in Fig. 2.10, an edge caching network composed by one macro BS and multiple cells is considered. The macro BS connects to the core network with wired links, while the connection between the macro BSs and small cells are wireless links.

Each cell has two states, active mode and sleeping mode. A UE can receive services from a cell only when it is within the coverage of the cell and the cell is in the active state. Moreover, a single UE can be served by multiple cells as long as the above conditions are met. To simplify the problem model, the situation that each UE connecting to multiple small cells acquiring a file is considered. If a UE can not receive all parts of the required file from small cells, it needs to require the remaining parts from macro BS. The total energy consumption of the system includes the consumption that maintains active small cells, when the UE acquires part of the file from small cells and when the UE acquires part of the file from the macro BS. Acquiring files from the macro cell needs more energy compared to that of acquiring files from the small cell. Therefore, we need to apply as many small cells as possible to receive

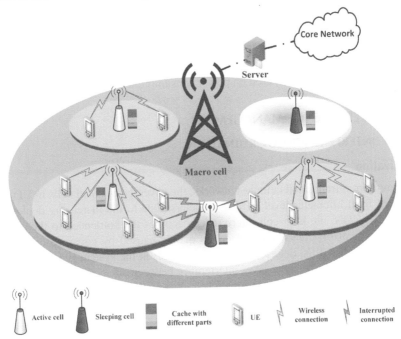

Figure 2.10 The application of energy-saving technique with edge caching in UDN

data while considering cell activation energy consumption to minimize total energy consumption.

To achieve the optimal edge caching network, the particle swarm optimization algorithm is used to find the optimal solution of the sleeping state of all the small cells B and the divided condition of all the files M. Specifically, the optimization algorithm contains four steps:

- Initialize the parameters used by the algorithm, including the initial B and M
- Calculate the value of ED by current parameters, and find the current optimal solution
- Update the velocity and location of the particles and start next iteration
- Find the global optimal solution after maximum iteration times

After deploying small cells with the appropriate strategy of content distribution and cell sleeping, UEs can acquire files from small cells. Each small cell sends its stored part of the file to UEs it covers. If a UE does not get the complete file, the macro BS retrieves the rest of file from the core network, then relays this to the relevant small cell that subsequently transmits this to the UE through the wireless link.

The proposed algorithm takes advantage of cell sleeping and content caching to improve energy saving and average hit rate. However, with the continuous development of communication network, traditional cell sleeping strategies are no longer sufficient for systems deployed with high density BSs. Thus, cell sleeping in BS collaboration methods should be taken into consideration, such as the collaboration between macro BS and small BSs [94], or the collaboration between adjacent BSs [95]. In addition, different sleep modes need to be considered, for example, different sleeping modes can be adapted based on power consumption levels [89].

2.3 INTERFERENCE MANAGEMENT

2.3.1 INTERFERENCE COORDINATION

Motivation

Different from traditional cellular networks, the coverage of small cells in UDN has reduced from kilometer level to tens of meters. Due to the increasing deployment density in UDN, the inter-cell interference becomes the bottleneck to the successful deployment of UDNs. Interference is mainly caused by coverage overlapping of different small BSs, deteriorating QoS and QoE in the system [96]. Interference can be divided into the same-layer interference between adjacent small BSs, and the cross-layer interference between macro BSs and small BSs.

- Same-layer interference
  The same-layer interference includes interference between macro cells and interference between small cells. The same-layer interference is mainly caused by the scarcity of spectrum resources, and it occurs when adjacent BSs use the same frequency band. Before the emergence of UDN, same-layer interference between macro BSs has already existed [97]. The study of interference management promotes the development of inter-cell interference coordination and CoMP. In UDN, most of the schemes about macro BS interference coordination cannot be applied to small BSs directly, due to the large number of small BSs and their irregular deployment. In general, BSs in the same layer have similar transmit power, so the interference problem in the same-layer is easy to solve compared with the cross-layer interference [98].
- Cross-layer interference
  Cross-layer interference mainly refers to the interference between the macro BS and the small BSs. Cross-layer interference includes the interference caused by the macro BS to the UEs in the small BSs and the interference caused by small BS to the UEs in the macro BS. The scarcity of spectrum resources and the difference of BS transmit power are both main factors leading to cross-layer interference. On the one hand, due to the strong path loss and attenuation, small BSs bring serious interference to the UEs at the cell-edge of the macro BS when small BSs reuse the frequency band of the macro BS. On the other hand, the macro BS needs to increase transmit power to ensure communication with the edge UE, which will generate interference to UEs in small cells [99]. At present, cross-layer interference has become a hot topic of research. However, the problem of

cross-layer interference is challenging because the transmit power of the macro BS is different from that of the small BSs [98].

Interference Coordination Method in UDN

The inter-cell interference would greatly reduce the network throughput. In order to prevent the serious decline of QoS, interference coordination methods in the network have been a major concern. As shown in Fig. 2.11, the existing interference coordination methods can be divided into time-domain interference coordination, frequency-domain interference coordination and spatial interference coordination.

- Time-domain interference coordination
  The time-domain interference coordination enables adjacent interfering BS to achieve orthogonality in the time domain. A commonly used interference coordination scheme is almost blank subframe (ABS) [100]. In ABS, the interference cell only emits ABS subframes in the time slot when the interference occurs, which only contains the basic control signal instead of the data signal. Moreover, the combination of cell collaboration and power control is typically exploited for time-domain interference coordination. Cell collaboration can improve spectrum efficiency and reduce network overhead.
- Frequency-domain interference coordination
  The frequency-domain interference coordination is to alleviate the signal interference by orthogonalizing the control signal and the physical signal of different cell. Fractional frequency reuse is a common frequency-domain interference coordination scheme [101]. In fractional frequency

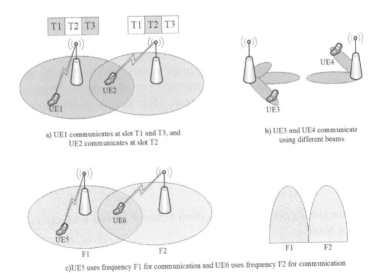

a) UE1 communicates at slot T1 and T3, and
UE2 communicates at slot T2

b) UE3 and UE4 communicate
using different beams

c)UE5 uses frequency F1 for communication and UE6 uses frequency F2 for communication

Figure 2.11   Different interference coordination methods

reuse, the entire frequency band is divided into orthogonal subbands and non-orthogonal subbands. UEs at the edge of the cell use the subband that is orthogonal to the adjacent cell, while UEs in the center of the cell use the non-orthogonal band, thus avoiding the interference to edge UEs. The subband partition scheme can use static and dynamic detection. Moreover, carrier aggregation can also be used to mitigate interference in the frequency domain [102]. The spectrum is stratified by two component carriers, one for reliable physical signal transmission and the other for data signal orthogonal transmission. Spectrum arrangement is also an efficient way to reduce interference [103], and beamforming codebook restriction is a typical scheme for spectrum arrangement. In the beamforming codebook restriction scheme, the UE can identify the channel and select the channel with good communication condition, thus alleviating the interference.

- Spatial interference coordination

Spatial interference coordination is mainly applied to eliminate spatial interference through highly oriented channel vectors [104]. Beamforming is an advanced interference coordination scheme [105]. By using beamforming, the macro BS selects the beam subset according to the number of UEs and the BS intensity. An appropriate beam subset selection can alleviate the cross-layer interference and improve the throughput. For UEs, using advanced antennas with interference suppression and elimination is also an effective solution [106]. By using digital signal processing, UEs can explore the interference source to choose an ideal communication direction, which greatly reduces the interference.

Spatial interference coordination is mainly applied to eliminate spatial interference through highly oriented channel vectors. That is, to transmit energy only in some directions and to create transmission opportunities for small cells in other directions [104]. Beamforming is an advanced interference coordination scheme [105]. By using beamforming, the macro BS selects the beam subset according to the number of UEs and the small BS intensity. The best beam subset selection can alleviate the cross-layer interference and improve the throughput. More details on beamforming are discussed in the later section. On the UE side, using advanced antennas with interference suppression and elimination is also an effective solution [106]. By using digital signal processing (DSP), UEs can explore the interference source to choose an ideal communication direction, which greatly reduces the interference.

Time-domain Interference Coordination Scheme Based on Generalized Interference Model

Due to the dense distribution of small cells in UDN, the same-layer interference occurs between different small cells. In order to reduce the interlayer interference between small cells, Yao *et al.* [107] propose a time-domain interference coordination scheme based on the generalized interference model. They use game

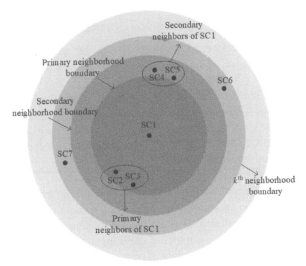

Figure 2.12 The proposed generalized interference model

theory to transform the time-domain allocation problem into a game. By solving the game problem, the existence of Nash equilibrium is proven. Therefore, according to the Nash equilibrium point, the optimal allocation scheme of ABS time slot is determined.

In order to minimize the interference of the whole network, a necessary method is to calculate the local interference of each BS and then accumulate the interferences into aggregated interference. In the process of calculating local interference, the traditional binary interference model only considers the interference caused by the neighbor nodes, which makes the result inaccurate. Therefore, they define a new interference model called the generalized interference model. Assuming that there are S small cells randomly distributed in an isolated cell, using $\mathscr{S} = \{1, 2, \ldots, S\}$ to represent the data set of all the small cells. Then, the generalized interference model is used. The generalized interference model takes UEs outside the traditional neighborhood into account and divides the neighborhood into k ($k \geq 2$) parts. Allocating different interference weights to neighbors with different distances. As shown in Fig. 2.12, in this network, $\mathscr{J}_1^0 = \{2, 3\}$, $\mathscr{J}_1^1 = \{4, 5\}$.

Then, the time slot division of ABS is defined. According to the generalized interference model, the local interference of node n can be represented. First, the ABS is divided into T equal length time slots, and the set of time slots is defined as $\mathscr{T} = \{1, 2, \ldots, T\}$, $(T < S)$. Then the time slot accessed by small cell n is defined as t_n, $t_n \in \mathscr{T}$. Finally, the interferences in all fields are accumulated, and the total interference can be obtained:

$$I_n = \sum_{m \in \mathscr{J}_n^0} \alpha_0 (d_{nm}) \cdot \delta (t_n, t_m) + \sum_{m \in \mathscr{J}_n^1} \alpha_1 (d_{nm}) \cdot \delta (t_n, t_m)$$
$$+ \cdots + \sum_{m \in \mathscr{J}_n^k} \alpha_k (d_{nm}) \cdot \delta (t_n, t_m) \tag{2.15}$$

where $\delta\left(t_m,t_n\right) = \begin{cases} 1, m = n \\ 0, m \neq n \end{cases}$ is an indicator function, which indicates whether the two nodes may cause interference, where $\alpha_0, \ldots, \alpha_k$ are different weight functions about d_{nm}. The interference of the whole network is defined as the accumulation of all local interferences, which is defined as $L_I = \sum_{n \in \mathscr{S}} I_n$.

Finally, the problem is solved using game theory where the BSs are the players, the choices to access ABS time slots are the strategies of the players, and the total network interference is the final benefit. The final payoff function is defined as $P1 : (t_1, \ldots, t_S) = \arg\min L_I$. Then, the best ABS slot allocation problem can be solved by finding the Nash equilibrium of the game.

In addition to interference coordination in time-domain, several coordination schemes are jointly considered. In frequency-domain interference coordination [108], enhanced dynamic cell muting is applied to interference coordination. CoMP transmission is used to realize a dynamic muting mechanism, which makes some BSs mute some resources, thus improving the transmission efficiency of other BSs while reducing interference. In spatial interference coordination [109], MIMO is also widely used. MIMO uses multiple antennas at the transmitter and receiver to make different signals propagate in different spatial positions, thus realizing spatial reuse and reducing interference.

2.3.2   SPECTRUM SHARING

Motivation

As one of the enabler technologies in 5G, UDN is expected to offer seamless communication, high capacity, and low delay. Toward this goal, the spectrum of 5G may be extended from 1 GHz to 100 GHz, already reaching millimeter level. For the future 6G, experts believe that the terahertz spectrum will be adopted. The scarcity of spectrum resources will become a bottleneck in the development of wireless communication technology. The shortage of spectrum resources is an important reason for inter-cell interference. Meanwhile, with the development of the IoT, a large number of IoT devices will be connected to the unlicensed industrial scientific medical bands, resulting in congestion in the industrial scientific medical band [110]. Multistandard network of traditional mobile operators occupy a certain amount of spectrum resources. As spectrum resources cannot be shared between different systems, serious wastage of spectrum resources exists. In order to solve the above problems, spectrum sharing technology has become the focus of current research.

Definition

Spectrum sharing enables the sharing of the original UE exclusive spectrum segments in a specific way, allowing two or more UEs to use the specified spectrum segments together. The UEs involved in spectrum sharing are generally divided into primary UEs and secondary UEs. The primary UEs refer to the UEs who are originally granted the frequency band and are willing to share the resources with other

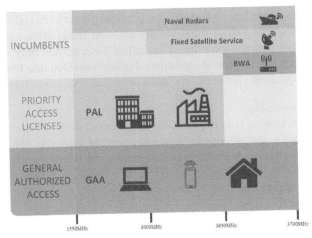

Figure 2.13 The schematic diagram of spectrum sharing

UEs. The secondary UEs refer to the remaining UEs who are allowed to follow the shared rule and use the spectrum. Spectrum sharing can dynamically allocate spectrum resources on demand, ensuring that the spectrum resources are not wasted. In addition, it can improve the spectrum efficiency, system capacity and the throughput of the cell while reducing energy consumption [111].

Figure 2.13 depicts an example of spectrum sharing in US, which was originally used for military radar and satellite equipment. In order to improve spectrum efficiency, the spectrum is shared with civilian use. Access is divided to different spectrum segments, among which intelligent factories and cities have priority access license and ordinary communication fields have general authorized access license. A UE with general authorized access license only uses idle spectrum without generating any interference to the normal communication of UEs with priority access license. With regard to the practical application of spectrum sharing, the effective combination of cognitive radio and spectrum sharing can be used to solve the problem of spectrum shortage [112]. Cognitive radio can improve the spectrum efficiency by adaptively changing the spectrum distribution according to historical experience. However, it is necessary to use an appropriate spectrum sharing scheme to reduce the interference. On the other hand, when multiple operators coexist, different operators have their own fixed frequency bands and different spectrum authorization systems. In order to achieve spectrum sharing among multiple operators while reducing interference, the regulatory bodies have provided new authorized spectrum methods to solve this problem. The following two aspects are introduced in detail.

Spectrum Sharing Scheme in Cognitive Radio

Cognitive radio [113] can realize multiplexing of the spectrum without changing the existing spectral resource allocation. The idea of cognitive radio is to use dynamic spectrum access technology, which divides UEs into primary UE and secondary UE. A cognitive radio network can automatically sense the usage of the surrounding frequency spectrum, and the secondary UEs utilize the spectrum only when primary users are unavailable. In general, spectrum sharing is divided into four steps: spectrum sensing, spectrum allocation, spectrum access, and spectrum handoff.

- Spectrum sensing
  Spectrum sensing is to detect available spectrum in the frequency band of time domain, spatial domain, frequency domain, and other spaces for primary UEs. Spectrum sensing can realize spectrum sharing while improving spectrum efficiency. It is the foundation of cognitive radio and the first step of spectrum sharing.
- Spectrum allocation
  The purpose of spectrum allocation is to allocate available spectrum to secondary UEs in a reasonable manner. Because the number of spectrum and secondary UEs can not be determined, an effective spectrum allocation algorithm is designed to allocate spectrum according to the QoS requirements of secondary UEs.
- Spectrum access
  The objective of spectrum access is to avoid conflict between primary UEs and secondary UEs. primary UEs have priority access to the allocated spectrum. In order to make secondary UEs available spectrum, it is necessary to design an efficient algorithm to coordinate the access of multiple secondary UEs.
- Spectrum handoff
  Spectrum handoff occurs in some special cases when secondary UEs switch to the new spectrum. There are generally three cases: the current spectrum of the secondary UE cannot meet the minimum QoS, primary UEs use this spectrum, and the position of the secondary UE moves out of the coverage.

Spectrum Sharing Schemes for Mobile Operators

Different countries and mobile operators have their own spectrum authorization systems which leads to various spectrum sharing schemes. The authorization system is generally formulated by the national spectrum regulatory agencies. The characteristics of authorization systems are mainly reflected in the degree of QoS, the level of spectrum access, the cost of frequency permission and the utilization of frequency bands. At present, the spectrum authorization system is roughly divided into three types [114]: individual authorization, light licensing, and general authorization.

- Individual authorization
  In individual authorization, a specific spectrum in a frequency band has specific access rights, and only licensed devices can access spectrums. These permits are generally issued by the national regulatory authorities of each country at a specific period. Therefore, the licensed UEs can use the frequency band unrestrictedly to meet the QoS requirements, but resulting in

spectrum wastage. In order to solve this problem, the existing spectrum sharing methods include co-primary shared access and licensed / authorized shared access.

- General authorization

  In general authorization, there is a frequency band for the exemption license. The UE can use the spectrum to achieve the sharing of the spectrum as long as the UE passes the authentication and complies with the general-defined technical regulations. The original authorized UE of the spectrum is referred to as a primary UE and a newly licensed UE is referred to as a secondary UE. The existing spectrum sharing methods include secondary horizontal shared access, unlicensed shared access, and unlicensed primary shared access.

A Application of Spectrum Sharing in Multi-operator Collocated UDNs

In UDN, it is difficult for operators to build a shared network and manage spectrum for densified cells. Georgios et al. [115] transformed the spectrum sharing problem into a network optimization problem, and proposed a spectrum sharing scheme to minimize the network density and spectrum sharing cost.

In this model, BSs are evenly distributed, and the distribution of BSs and its UEs follows a PPP distribution of λ_b and λ_u ($\lambda_b \gg \lambda_u$). Denote α as a parameter related to distance and ρ_0 is a constant as defined in [116], then the average UE capacity can be defined as:

$$Cp(\lambda_b, \alpha) \approx \log\left(1 + \left(\frac{\lambda_b}{\rho_0 \lambda_u}\right)^{\frac{\alpha}{2}}\right) \tag{2.16}$$

Then, the constraint that the UE can communicate normally can be defined. Denote W as a fixed spectrum, t_u is the average UE requirement of the UE using this spectrum, and the average subscriber rate that the network can provide to the UE is the product of W and $Cp(\lambda_b, \alpha)$. In order to achieve the communication requirement of the UE, the spectrum W and BS distribution λ_b of the operator are optimized to meet $t_u \leq WCp(\lambda_b, \alpha)$.

Then the optimization problem is defined under the constraint. Considering that $N > 1$ operators co-exist in a common deployment network, as shown in Fig. 2.14. For operators n, $\lambda_{d,n}$ as the BS distribution, $\lambda_{u,n}$ is the UE distribution, $t_{d,n}$ is the average service demand, W_{max} is the total spectrum sharing pool of all operators, W_n ($\sum_n W_n \leq W_{max}$) is the dedicated spectrum of operator n, $c_{b,n}$ is the cost of deploying BS, and $c_{w,n}$ is the cost of spectrum. In order to solve the optimal distribution of BS and spectrum distribution of each operator, the problem is modeled as follows:

$$\begin{aligned}
\text{minimize} \quad & \sum_{n=1}^{N} c_{b,n} \lambda_{b,n} + c_{w,n} W_n, \\
\text{subject to} \quad & t_{u,n} \leq W_n \log\left(1 + \left(\frac{\lambda_{b,n}}{\rho_n \lambda_{u,n}}\right)^{\frac{\alpha}{2}}\right) \quad n = 1, \ldots, N, \\
& \sum_{n=1}^{N} W_n \leq W_{max}, \\
& \lambda_{b,n} \leq \lambda_{b,n,max}
\end{aligned} \tag{2.17}$$

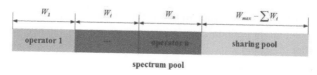

Figure 2.14 Multi-operator spectrum sharing method

Finally, the solution is to transform the above problems into a convex optimization problem, and the variables $\lambda_{b,n}$ and W_n are transformed into joint convex variables. However, as the above two variables can not be solved directly, it is necessary to use Lagrangian duality method to transform the optimization problem into dual problem. For the dual problem, convex optimization is used to update two dual variables iteratively until they converge, and the solution is the optimal solution.

In addition to the spectrum sharing strategy for multiple operators, multiple spectrum sharing schemes are jointly considered. In dynamic spectrum sharing [117], the dynamic spectrum distributor determines the distribution of the current spectrum through the past spectrum usage records, so as to achieve spectrum sharing. In the multi-dimensional spectrum sharing optimization based on polarization [118], not only the traditional dimension resources i.e., time, frequency, space, and power considered, but also the polarization resources are considered. By optimizing the multi-dimensional resources of multiple cells, the network capacity is maximized and the interference is minimized, so that spectrum sharing can be realized.

2.3.3 BEAMFORMING

Motivation

As most of the frequency bands below 3 GHz are occupied by 4G and other networks, mmWaves between 30 GHz and 300 GHz are considered in 5G. Although mmWave has broad frequency band, the receiving power is very low due to the short wavelength of mmWave. Moreover, the poor penetration ability of mmWave leads to serious signal attenuation, which increases the defection degree of mmWave transmission by environmental factors. Therefore, to improve transmission efficiency of mmWave and reduce signal interference, massive MIMO is used. Massive MIMO can effectively solve the interference problem in mmWave [119]. Although the environment interference is solved, the interference between mmWaves still exists when multiple UEs communicate at the same time. Beamforming has been applied and widely studied in 5G to steer the signal direction transmitted by the UE specified and reduce interference between the signals.

Definition

Terms beamforming and MIMO are sometimes used interchangeably. MIMO realizes spatial reuse by using multiple antennas at both the transmitter and the receiver. However, due to the fact that multiple antennas transmit signals at the same time,

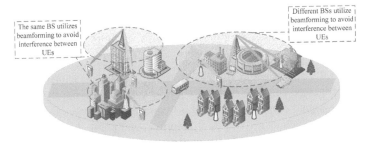

The same BS utilizes beamforming to avoid interference between UEs

Different BSs utilize beamforming to avoid interference between UEs

Figure 2.15 Schematic diagram of beamforming

and the distance of multiple antennas is very close, multiple signals easily collide with each other. This situation results in interference between signals, which greatly reduces communication quality and adversely affects transmit signals. Therefore, in order to solve the problems existing in MIMO system, beamforming is introduced. The main idea of beamforming is to make the wave propagate in a fixed direction and weaken in other directions based on the principle of the wave interference and diffraction. In order to realize the correct superposition of multiple waves, beamforming needs to adjust the amplitude and phase of the transmitted signal in each antenna so that the wave is superimposed in a specific direction. As shown in Fig. 2.15, digital beamforming has been used in 4G networks. Digital beamforming requires the configuration of independent RF links for each antenna [120], while it is too costly for 5G networks with massive MIMO. Therefore, for 5G beamforming, the hybrid beamforming based on digital beamforming and analog beamforming is used. The new hybrid beamforming algorithm has also become the focus of current research.

Hybrid Beamforming Structure

Beamforming needs to adjust the amplitude and the phase of signals transmitted by each antenna according to prior information, so that the superimposed signal can be transmitted in the specified direction. There are two traditional beamforming methods: baseband digital beamforming and analog beamforming. Pure digital beamforming uses a digital processor to process the signals to make beamforming flexible with more degrees of freedom and less internal interference. However, the complex hardware implementation of pure digital beamforming has high energy consumption, and each antenna should be equipped with a special radio frequency link, making the application of pure digital beamforming to massive MIMO system costly. Analog beamforming uses time extension unit or equivalent phase to process the signals, reducing flexibility and generating some internal interference. However, the hardware implementation of analog beamforming is simple, and needs less energy consumption and lower cost. Based on the pros and cons of analog beamforming and digital beamforming, hybrid beamforming is considered to be the most suitable technology for 5G networks [121]. At present, there are two kinds of hybrid beamforming structures: fully-connected hybrid beamforming and sub-connected hybrid beamforming.

- Fully-connected hybrid beamforming
  In the fully-connected hybrid beamforming system, the signal is first digitally processed and then analogally processed. After digital signal processing, the signal output of each radio frequency link passes through multiple devices such as mixer, power amplifier, and phase device. Then, these signals gathers in front of each generator and wait for further processing. Although such a complex structure is difficult to realize, it provides full-beamforming gain for each transmitter, and has better system performance.
- Sub-connected hybrid beamforming
  In the sub-connected hybrid beamforming system [122], the signal also needs to be digitally processed and analogally processed. The difference is that each RF link only needs to be connected to part of the antenna after the process of mixing, amplification, and displacement, which means each generator is only connected to one RF link. This structure greatly reduces complexity and energy consumption, so it can be conveniently deployed in a handheld equipment with low power. However, it reduces beamforming gain and system performance.

An Application of Full-connected Hybrid Beamforming in UDN

In order to reduce system complexity as much as possible, designing hybrid beamforming schemes have become the focus of current research. Sohrabi *et al.* proposed a fully connected hybrid beamforming scheme [123], in which the hybrid beamforming performance is close to the pure digital beamforming. Using the heuristic method, the minimum number of RF chains used by the system can be obtained, and the cost of the whole system can be reduced.

First of all, the model of MIMO system can be established by considering the single UE MIMO system, as shown in Fig. 2.16. The number of antennas at the transmitter is N, the number of antennas at the receiver is M, and the transmitted data stream is N_s. Then the number of RF chains required by the transmitter is

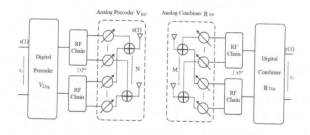

Figure 2.16 A massive MIMO system with hybrid beamforming structure at the transmitter and the receiver

N_t^{RF} ($N_s < N_t^{RF} < N$), and the RF chain required by the receiver is N_r^{RF} ($N_s < N_r^{RF} < M$). In general, $N_t^{RF} << N$ and $N_r^{RF} << M$. In the specific hybrid beamforming structure, considering the transmitter, the RF precoder for analog signal processing is $\mathbf{V}_{RF}$ ($\mathbf{V}_{RF} \in \mathbb{C}^{N \times N_t^{RF}}$), and the precoder for the digital signal processing is $\mathbf{V}_{Dig}$ ($\mathbf{V}_{Dig} \in \mathbb{C}^{N_t^{RF} \times N_s}$). Then the hybrid precoder $\mathbf{V}_t = \mathbf{V}_{RF}\mathbf{V}_{Dig}$. Similarly, at the receiver, the RF combiner for analog signal processing is $\mathbf{K}_{RF}$ ($\mathbf{K}_{RF} \in \mathbb{C}^{M \times N_r^{RF}}$), and the combiner of digital signal processing is $\mathbf{K}_{Dig}$ ($\mathbf{K}_{Dig} \in \mathbb{C}^{N_r^{RF} \times N_s}$), hybrid combiner $\mathbf{K}_t = \mathbf{K}_{RF}\mathbf{K}_{Dig}$.

Then, assume the analog phase shifters are used to implement the elements of RF beamforming, so the elements of the RF deployment forming matrix are constrained to $|\mathbf{V}_{RF}(i,j)|^2 = 1$ and $|\mathbf{K}_{RF}(i,j)|^2 = 1$. As shown in Fig. 2.16, in hybrid beamforming, considering the transmitter, the linear transmit precoded signal can be defined as $\mathbf{x} = \mathbf{V}_{RF}\mathbf{V}_{Dig}\mathbf{s}$, where $\mathbf{s}$ is a vector of N_s signal. Then considering the receiver, $\mathbf{z}$ is white Gaussian noise, and the processing signal at the receiver after using the combiner can be obtained as $\mathbf{y} = \mathbf{K}_{Dig}^H\mathbf{K}_{RF}^H\mathbf{H}_{RF}\mathbf{V}_{RF}\mathbf{V}_{Dig}\mathbf{s} + \mathbf{K}_{Dig}^H\mathbf{K}_{RF}^H\mathbf{z}$. Finally, denote R is the transmission rate for the entire system as $R = \log_2\left|\mathbf{I}_M + \frac{1}{\sigma^2}\mathbf{K}_t\left(\mathbf{K}_t^H\mathbf{K}_t\right)^{-1}\mathbf{K}_t^H\mathbf{H}_{RF}\mathbf{V}_t\mathbf{V}_t^H\mathbf{H}_{RF}^H\right|$, where σ^2 is the variance of white Gaussian noise.

The system aims to maximize the transmission rate satisfying the total transmit power and the RF matrix constraints, which can be defined as the following optimization problem:

$$\begin{aligned} \underset{\mathbf{V}_{RF}, \mathbf{V}_{Dig}, \mathbf{K}_{RF}, \mathbf{K}_{Dig}}{\text{maximize}} \quad & R \\ \text{subject to} \quad & \text{Tr}\left(\mathbf{V}_{RF}\mathbf{V}_{Dig}\mathbf{V}_{Dig}^H\mathbf{V}_{RF}^H\right) \leq P, \\ & |\mathbf{V}_{RF}(i,j)|^2 = 1, \forall i,j \\ & |\mathbf{K}_{RF}(i,j)|^2 = 1, \quad \forall i,j \end{aligned} \qquad (2.18)$$

where P is the total transmit power allocated at the transmitter. Before solving the optimization problem, it should be proved that optimal solution can be obtained in the case of $N_t^{RF} > 2N_s$. Then, because of its non-convexity, a heuristic method is proposed to solve the optimization problem. Finally, through the optimal solution, the optimal transmit precoder and receiver combiner are designed, and the optimal hybrid beamforming scheme is assumed.

In addition to the fully-connected hybrid beamforming scheme, several beamforming schemes are jointly considered. In the location based beamforming scheme [124], the line-of-sight path arrival angle between the access node and the UE node is considered, and the extended Kalman filter method is used to track and estimate the beamforming weight vector at the receiver and transmitter. In the hybrid beamforming scheme based on partial CSI [125], the spatial covariance matrix can be estimated by using the spatial construction algorithm of partial CSI, and a unified analog beamforming algorithm is designed according to the matrix information.

2.4 SUMMARY

In this section, we first introduce two common modeling techniques in UDN, stochastic geometry and game theory. The basic idea of stochastic geometry is to assume that the positions of nodes or network structures are randomly distributed, so that the network topology can be established using a series of random PP, such as PPP, BPP, PCP, and HCPP. The stochastic geometry model fully considers the unpredictability of the nodes in the wireless network, and can provide support for network mode selection, AP association, and state control in heterogeneous networks. Game theory is a mathematical model that studies the interactions between individuals to resolve conflicts between a group of rational entities. Through the game theory model, a system can more comprehensively analyze the interactive decision-making process between entities, and provide solutions for complex distributed optimization problems. In UDN, game theory model can effectively solve problems such as interference management, resource allocation, small cell discovery, and UE association, which maximizes the benefits of each UE while ensuring fairness.

Then, we analyze three common techniques in resource management, such as NOMA-based QoE provisioning technique, load balancing, and energy-saving technique. Compared with OFDMA, the application of NOMA can improve system spectral efficiency as well as ensure fairness between UEs while meeting the throughput requirements. We focus on discussing the CRE technique in load balancing, although MDP-based solution has advantages in accurate prediction and game theory-based solution has advantages in global optimization. The CRE technique is suitable for inter-layer equalization schemes, which can reduce uplink interference while being simple to implement. Energy saving technologies are mainly targeted at BSs: energy-harvesting technique is considered to save energy with renewable energy, cell-deployment technique is considered to save energy through reasonable overall structure, cell-sleeping technique is considered to save energy by turning off some idle BSs. Among the above technologies, cell sleeping is the most effective way to improve network energy efficiency and reduce energy waste.

Finally, we introduced three common techniques in interference management, including interference coordination, spectrum sharing, and beamforming. The aim of interference coordination is to reduce the same-layer interference between neighboring small BSs and the cross-layer interference between macro BS and the small BSs. The current interference coordination technologies can be divided into four types, including time-domain, frequency-domain, spatial and power control interference coordination. Time-domain and frequency-domain interference coordination are achieved by making signals orthogonal in different resource domains, spatial interference coordination is achieved by using highly directional channel vectors to eliminate spatial interference, and power control interference coordination is achieved by adjusting the power of the transmitted signals, which improves QoE of UEs in different cells. Spectrum sharing technique is mainly used to solve the problem of spectrum scarcity and spectrum resource wastage, which is a hot spot in current research. In cognitive radio, spectrum sharing is generally divided into four steps: spectrum sensing, spectrum allocation, spectrum access, and spectrum handover. However, spectrum authorization systems between different operators have large differences, wherein different authorization systems have their own spectrum sharing schemes. Therefore, it is necessary to fully consider the differences of multiple operators in the design of spectrum sharing strategy to optimize costs. The main idea of beamforming technique is to propagate waves in a fixed direction through the principle of wave interference and diffraction. Beamforming techinque is an evolution of MIMO, and solves the problems of large interference and low communication quality in MIMO systems. Hybrid beamforming is currently the most suitable technology for 5G networks, it can be further divided into fully-connected hybrid beamforming and sub-connected hybrid beamforming, which can reduce the cost of equipment while making beamforming have more freedom and less errors.

3 Mobility Management

3.1 UE ASSOCIATION

3.1.1 UE ASSOCIATION IN CONVENTIONAL CELLULAR NETWORKS

Motivation for UE Association

UE association (or cell association) refers to the process of mapping one or several BSs to serve a given UE. Driven by immense traffic, heterogenous UDN is featured by small and irregular cells, busy backhauling, and imbalanced user distribution over small cells. The traditional UE association methods enable a large number of users to access the macro station, resulting in a serious imbalance between the load of the macro BS and the micro BS, and the wireless resources cannot be effectively utilized. Therefore, how to develop an effective association mechanism to balance BS load and control traffic will become more intractable than before [126].

Existing UE Association Algorithms for UDN

In this section, we will introduce the UE association algorithms (as shown in Fig. 3.1) from the perspective of associated metrics, challenges, and common research models.

UE Association Metrics

The metrics for measuring UE association performance mainly include the following aspects: outage/coverage probability, spectrum efficiency, energy efficiency, QoS, and fairness.

- Outage/Coverage probability
  The probability that the SINR falls below a certain threshold is defined as outage probability. On the contrary, coverage probability refers to the probability that the SINR rises above a certain threshold. The outage probability characterizes the link capacity. When the link capacity cannot meet the data rate required by the user, an outage event occurs. This event is probabilistic and depends on the link's average signal-to-noise ratio and its channel fading distribution model. Therefore, outage probability is an important indicator of link capacity.
- Spectrum efficiency
  When measuring the efficiency of a communication system, attention should be paid not only to the transmission rate, but also to the bandwidth of the channel required under such a transmission rate. The symbol transmission rate in the unit band measures the transmission efficiency of

DOI: 10.1201/9781003148654-3

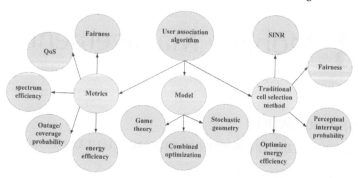

Figure 3.1 Classification of UE association algorithm

a communication system, that is, the spectrum efficiency. Therefore, some UE association schemes are designed to optimize system spectrum efficiency to serve as many users as possible.

- Energy efficiency

  In recent years, researchers pay increasing attention to green communication technology to maintain environmental sustainability. In UDNs, the deployment density of small BSs is very high, and energy consumption has become a factor that cannot be ignored. Therefore, while considering the capacity improvement brought by small BSs, the energy efficiency of the network system in UDN becomes a major concern of designing algorithm.

- QoS

  QoS metric can be effectively quantified by transmission delay, UE throughput, and SINR. Since different BSs provide different signal strength, channel quality, and available resources for each UE, the UE association strategy needs to consider the QoS requirements of the UE. Hence, QoS is also an important indicator for operators and customers.

- Fairness

  In the traditional cellular network, wireless resources are allocated to UEs mainly through the resource scheduling by the macro BS. In UDN, fairness issues occur not only in scheduling, but also in UE association decisions between cells at different levels. Fairness is mainly reflected in providing more resources for UE with low channel quality and encouraging them to associate with BSs with more resources. The Jain's fairness index is widely adopted to evaluate the fairness, which is defined as

$$F_J(H_1...,H_n,...H_N) = \frac{\left(\sum_{n=1}^N H_n\right)^2}{N\sum_{n=1}^N H_n^2} \qquad (3.1)$$

where N is the number of UE and H_n is the throughput of the nth user device.

In addition, when designing a UE association strategy, other resource management mechanisms in the network should also be considered, such as interference management, resource allocation, power control, and energy management, to provide network service quality assurance for the UE while improving system performance.

Challenges of UE Association

As the deployment of BSs becomes densification, UE association faces more challenges than traditional cellular networks. For example, issues such as handover control, services and management related to UE association, interference, have presented new characteristics in UDN environment.

- Handover control
  An ideal handover control mechanism enables UEs to seamlessly handover between different types of cells in UDN. However, due to the irregular coverage and the complicated BS neighborhood relationship, UEs may experience problems such as frequent handover and ping-pong effects during their movement. Moreover, different types of BSs will lead to complicated handover signaling procedures. Therefore, the mobility management mechanism needs to be redesigned to adapt to such complicated signaling handover.
- Asymmetric downlink and uplink
  In the downlink, due to the power differences between different types of BSs, the coverage of macro cells is much larger than that of small cells. In contrast, UEs shield the heterogeneity of BSs. A UE can send the same power level over an uplink regardless of BS type. In addition, the downlink traffic is usually much larger than the uplink traffic. Based on these asymmetries, the optimal downlink association is necessarily not the optimal uplink transmission. Therefore, joint load balancing studies are necessary for the downlink and uplink.
- Interference
  Ultra-dense environments with a large number of BSs have brought more interference sources. For example, in a crowded subway, there are a large number of UEs and micro BSs. Therefore, the signal may have more reflection and scattering paths. In this interference environment, how to select BSs for association is also a key issue to be solved.

Therefore, the UE association in UDN needs to consider an amorphous, dynamic, and virtual location area, which is formed by BSs and follows the users' movement.

Common Research Models

- Game theory model
  In the game theory model, users change their antagonistic strategies according to the opponent's strategy to win the game. Game theory considers the predictive and actual behavior of individuals in the game, and studies their

optimization strategies to optimize the performance of the entire system. Therefore, the focus of the game theory model is to characterize the competitive relationship between users. A detailed introduction to game theory is provided in Section 2.1.2, and we will not go into details here.

- Combined optimization model

In the combined optimization model, user requests and BS resources are jointly considered to maximize system availability under the constraints of limited resources. In this model, the association between UEs and BSs is expressed by a variable. By constructing a utility function using the association variables as independent variables, the UE association problem is transformed into a utility maximization problem. Therefore, the combined optimization model focuses on the cooperative relationship between users and BSs.

- Stochastic geometry

In UDN, the heterogeneity of the network and the introduction of new technologies lead to an increase in terms of network complexity. The Wiener model commonly used in traditional cellular networks is unable to accurately analyze this complex cellular network architecture. Therefore, as a statistical tool, stochastic geometry is introduced into the communication system. The main idea of stochastic geometry is to abstract the stochastic position of nodes into a suitable stochastic point process. Stochastic geometry method can effectively describe the randomness of the wireless communication network topology, and accurately analyze the network performance theoretically. The stochastic geometry has been introduced in detail in Section 2.1.1.

A Cellular System with UE Association

BS sleeping has become a viable solution to improve overall network energy efficiency by deactivating underutilized BSs. However, it affects the quality of service and user experience of UEs in the sleeping cell. In this subsection, we introduce an explanatory example to describe how the UEs are associated during the BS sleeping. This example refers to the UE association architecture in [127].

Figure 3.2 shows a downlink network with L circular macrocells. The UEs are evenly distributed in L cells. In this example, $L = 7$ and BS 1 is the sleeping BS. The basic idea is as follows: according to the signal strength distribution and signal power distribution received by UEs in BS 1 cell, the probability density function, the cumulative distribution function and the moment generating function are gained by the Cartermont approximation method. Then, these three functions are used to calculate the maximum mean channel access probability. The maximum mean channel access probability is defined as the maximum value of the probability that any user in a sleeping cell will obtain a transmission channel from all active BSs. Any user in sleeping cell BS 1 selects the BS (among all active BSs) that provides a maximum chance to obtain a channel, that is, the BS that provides the maximum mean channel access probability. For example, according to the above calculation results, there is

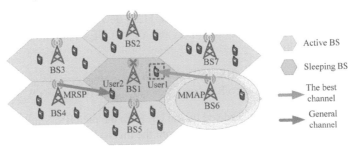

Figure 3.2 Graphical illustration of the macro cell network with different BSs in sleeping modes

a maximum mean channel access probability between User 1 and BS 6. User 1 and thus will be connected to BS 6.

3.1.2 CELL RANGE EXPANSION (CRE) IN UDNS

Motivation for Cell Range Expansion in UDNs

In UE association mechanisms, a macro BS with high transmit power will be used as a serving cell for most UEs. Thus a UE will not choose to access the pico BS, even if the pico BS is nearest to the UE. As a result, the number of UEs covered by the macro BS far exceeds that of the pico BS, which will inevitably lead to the overload of the macro BS and the resource waste of the pico BS. Ultimately, the inter-cell traffic load is unbalanced, and the limited spectrum resources can not be effectively utilized. In addition, when the transmission power of all BSs in the cellular network is the same, the uplink and downlink can choose the same service cell with the same coverage. However, the coverage of the pico BS is smaller than that of the macro BS, and thus, UEs are likely to choose macro BS as the serving cell, which can also cause the resources of the pico BS to be underutilized and the macro BS to be overwhelmed. In contrast, the CRE cell selection process achieves load balancing and spatial reuse by connecting UEs to a nearby pico BS, and mitigates inter-cell interference by reducing the transmit power of UEs.

The Expanded Region of Pico Cells in Heterogeneous Networks

CRE adds positive range extension bias to the received signal strength of the pico cell downlink at the UE to increase the downlink coverage of the pico BS. Therefore, how to set the value of range extension bias is a challenge. Specifically, if the bias is set too small, there are fewer UEs connecting to the pico BS, and the resources of the pico BS may be wasted, which may also make the macro BS to be overloaded. If the bias is set too large, the resources of the pico BS are limited, and the number of UEs connecting to the pico BS is large, which may degrade the QoS of some UEs, and the resources of the macro BS are not rationally utilized.

To resolve this contradiction, David *et al.* [128] developed an OFDMA network comprising a macro BS, a set of pico BSs and a set of UEs. And the selection rules of the BS are as follows:

- Rules for selecting pico BS: $(\omega_p)_{dB} + (\Delta B)_{dB} > (\omega_m)_{dB} + (Q)_{dB}$
- Rules for selecting macro BS: $(\omega_m)_{dB} > (\omega_p)_{dB} + (\Delta B)_{dB} + (Q)_{dB}$

$(\omega_p)_{dB}$ is the signal strength received by the UE from the pico BS. Similarly, $(\omega_m)_{dB}$ is the signal strength received by the UE from the macro BS. $(\Delta B)_{dB}$ is the bias and $(Q)_{dB}$ is a hysteresis margin. As can be seen from the above inequalities, bias $(\Delta B)_{dB}$ plays a key role in CRE. During the implementation of the cell range expansion algorithm, there are two main boundary concepts that need to be defined and bias $(\Delta B)_{dB}$ is added just for the purpose of extending the service boundary of the pico cell.

- The equal received signal strength (RSS) boundary (ESB)
  ESB refers to the boundary where the signal strength received by a UE from a pico BS is equal to that received from a macro BS. The red circle in Fig. 3.3 indicates the ESB.
- The equal path loss boundary (EPB)
  EPB is the boundary where the path loss experienced by a UE when connecting to a pico BS is the same as that when connecting to a macro BS. Any UE on the green circle in Fig. 3.3 has the same path loss whether it is connected to the macro BS or the nearest pico BS.

The traditional UE association rule is based on ESB, that is, the UE is connected to a specific BS according to the received signal strength. This association rule only allows UEs in the red circle to connect to the pico BS, i.e., the ESB is the coverage

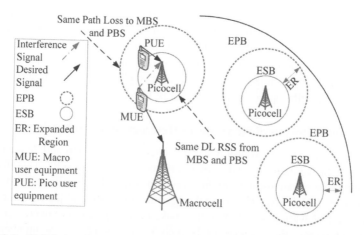

Figure 3.3 Uplink interference scenarios in macrocell–picocell deployments. The equal Downlink RSS boundary (ESB) and the equal path-loss boundary (EPB) of a picocell as defined in article are also plotted

boundary of the pico cell. CRE can utlize the path loss as the basis for setting the $(\Delta B)_{dB}$, and extend the coverage boundary of the pico cell from ESB to EPB, so that more UEs can be connected to the pico BS.

3.1.3   COORDINATED MULTIPLE POINTS IN UDNS

Motivation

As aforementioned, small BSs are densely deployed in UDN, and a UE may be in the coverage of multiple small cells. If the UE selects only one BS to connect, the signals of other BSs are interference to the UE. In addition, in the 3GPP long term evolution-advanced (LTE-A), neighboring cells completely multiplex frequency resources with a frequency reuse factor of 1 [129]. Although this strategy can significantly improve the system capacity, the complete reuse of frequency resources will cause serious inter-cell interference, which affects the service quality of UEs at the cell edge. In this case, a CoMP technique is introduced. CoMP refers to a technology in which multiple geographically separated BSs cooperatively participate in serving one UE. This technology can effectively eliminate interference by sharing some necessary information between BSs.

CoMP Technology

The core technologies of CoMP are mainly classified as the following two categories:

- Joint transmission/Joint reception
  When connected through a large capacity, non-delayed backhaul network, the cooperating BSs can share not only the CSIs, but also UEs' data received by them. As shown in Fig. 3.4(a) and Fig. 3.4(b), the cells participating in the cooperation jointly transmit UE data, which can effectively improve UEs' receiving SINR and suppress inter-cell interference [130].
- Coordinated scheduling/Coordinated beamforming
  CSI enables cooperating BSs to adjust frequency and time, as well as co-ordinated beamforming direction and power allocation, thereby reducing cell-edge interference. Therefore, CSI sharing between BSs can improve network performance. As shown in Fig. 3.4(c) and Fig. 3.4(d), since there is no need to share the received data of UEs, the backhaul network bandwidth required for this collaboration is reduced.

Cooperative BS Selection Scheme

The cooperative BS set selection scheme mainly includes the following three parts:

- Static collaboration set partition
  The BSs in the network are grouped in the network initialization phase. The choice of this cooperative BSs is independent, and the BS set is static. Each UE can only be served by the uniquely determined BS set. The solution has

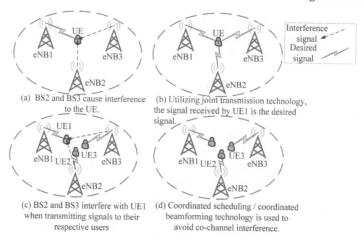

(a) BS2 and BS3 cause interference to the UE.

(b) Utilizing joint transmission technology, the signal received by UE1 is the desired signal.

(c) BS2 and BS3 interfere with UE1 when transmitting signals to their respective users

(d) Coordinated scheduling / coordinated beamforming technology is used to avoid co-channel interference.

Figure 3.4 Figure (a)(b) utilizes the Joint Transmission technology, which allows the three BSs in the cooperative set to transmit the same data to the UE using the same time-frequency resource, and the signals received by the UE are desired signals. Figure (c)(d) utilizes Coordinated Scheduling/Coordinated Beamforming technology, which can maximize the scheduling of corresponding mobile clients on different time-frequency resource blocks to avoid co-channel interference

the advantages of low complexity and low overhead, and does not require high capacity for the backhaul link [131]. However, the shortcoming of this scheme is that the operation is not flexible, and the interference of the neighboring BSs may not be effectively eliminated.

- Dynamic collaboration set partition
  The coordinated BS set is determined according to mobile UEs' needs, and the different coordinated BS sets may overlap. This scheme is applicable to the case of UE movement. When a UE moves, the scheme dynamically changes the composition of the UE collaboration set [132]. Therefore, this scheme is required to deal with resource allocation conflicts caused by the overlapping of cooperative BS set. However, the scheduling complexity of the scheme is very high, and UEs need to periodically measure and report the channel status of the surrounding BSs, and thus, resulting in high feedback overhead.

- Semi-static state cooperative set partition
  This scheme first statically sets a group of neighboring BSs for the UE, and then dynamically selects some suitable BSs from the neighboring BSs to form a set of coordinated BSs. The scheme is simple and easy to implement, but the downside is that the performance fluctuates greatly when the user is moving erratically and quickly.

3.2 MOBILITY MANAGEMENT

Motivation

The increased cell densification encountered in UDN continuously poses challenges for mobility support. The reduced transmit powers of small cells lead to reduced footprints. As a result, for a mobile user, a user association algorithm that does not consider the mobility issues may result in more frequent handovers among the cells in UDN compared to conventional cellular networks. However, it is well-understood that handovers impose costly overheads as well as undesirable handover delays and possibly dropped calls. Hence, it is imperative to account for user mobility when making user association decisions in UDN to enhance the long-term system-level performance and to avoid excessive handovers. In this section, we will describe mobility from the perspective of power control and handover management.

Criteria for Power Control

Power control technology in communication systems has been well studied. In principle, it can be divided into power control based on power balance criterion, SINR balance criterion, hybrid balance criterion and bit error rate balance criterion.

- Power balance criterion
  Power balance means that the signals received by the receiver are equal [133]. In the uplink, the power of each arriving signal is equal; in the downlink, if signals are sent by the same BS, the power should be controlled according to the distance between BS and each UE so that the power received by each UE is the same.
- SINR balance criterion
  SINR balance means that the SINR values received by the BS should be equal. When UEs arrive at the BS with equal power, their corresponding SINR is also equal, so it is equivalent to the power balance criterion. But for multi-cell UEs, the received signal strength will be different because of the interference.
- Hybrid power control criterion
  A hybrid criterion of power balance and SINR balance can be employed. For high-interference UEs, power balance criterion can be utilized, and weakly interfering UEs utilize SINR balance criterion.
- Bit error rate balance criterion
  In digital communication systems, the bit error rate is a measure of communication quality. Power control technology based on bit error rate balance is proposed. However, the bit error rate criterion has certain limitations, because the general bit error rate refers to the average bit error rate, i.e., it is the statistical average over time [134]. Therefore, the bit error rate criterion would cause delay, which would affect the accuracy of the entire power control process.

Common Methods for Power Control

There are many classification methods for power control. According to the way of obtaining power control information, it can be divided: open loop, closed loop, and outer loop power control [135].

- Open loop power control
  Open-loop power control is a widely used power control method [136]. It is called open-loop because it does not require a feedback loop and is relatively simple to implement. Open-loop power control provides a dedicated pilot channel for channel estimation, which is transmitted by the BS to all UEs. After receiving the pilot channel information, each UE estimates the power strength. Then the UE adjusts the transmit power accordingly. In open loop power control, the uplink and downlink are assumed to be related.

- Closed-loop power control
  The closed-loop power control is that the node adjusts its uplink transmit power according to the transmission power control. The transmission power control command is based on a preset target function value and a measured reception target function value, it is transmitted by the serving node to the served node [137].
  In a closed loop power control system, a serving node estimates the value of the received objective function. The target function value is compared with a predetermined objective function value, and the transmission power of the node is adjusted by the result of the comparison. As shown in Fig. 3.5, the closed-loop power control scheme works near the open loop point, first it adopts an open-loop power control scheme, and the node initially sets its transmit power; then, it adjusts its transmit power by receiving transmission power control commands from its serving node.

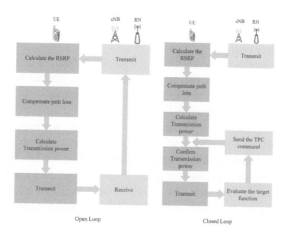

Figure 3.5   Closed Loop and Open Loop

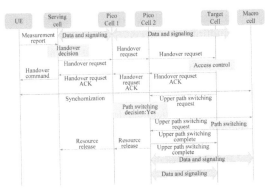

Figure 3.6   The handover process of a UE in pico BSs under the control of a macro BS

- Outer loop power control
  The outer loop power control sets specific target channel conditions based on the required target block bit error rate and sets the target signal-to-interference ratio based on the conditions. The outer loop power control allows for predefined QoS for each link, which reduces resource occupation and interference, thus increasing system-wide throughput.

Requirements for Handover

Mobility management requires that once the UEs of the network are unable to receive better communication services during the movement, UEs need to ensure continuity of communication services by switching to other links. As a representative technology of mobile communication networks, cell handover provides technical support for mobility management. As shown in Fig. 3.6, the handover process of a UE in pico BSs under the control of a macro BS. In cellular communications, the terms handover or handoff refer to the process of transferring an ongoing call or data session from one channel connected to the core network to another channel [138].

Common Handover Methods

During the handover process, it is necessary to ensure that the link transfer is seamless [139]. Especially in UDN, handover is the key to achieve seamless coverage of the communication network. According to the existing literature, the handover can be classified as follows:

- Hard/Soft handover
  Handover can be divided into hard handover and soft handover. Hard handover means "disconnect first, then connect". The UE disconnects with the current serving BS first, and then establishes a connection with the target BS. Hard handover can save channel resources, but "disconnect" may lead

temporarily disconnect from the current BS, and the "release-build" process may cause multiple handovers. Soft handover is "connect first, then disconnect" [140]. That is, after establishing a connection with the target BS, the connection with the current BS is disconnected. It can overcome the shortcoming of the hard handover interruption.

- Horizontal/Vertical handover
Horizontal handover and vertical handover are defined according to access technology of network. The handover of UE between isomorphic networks using the same access technology is called horizontal handover. Horizontal handover generally considers the RSS received by the UE and the remaining system resources of the target cell, which is relatively simple. Handover between different access technologies is called vertical handover, also called inter-system handover. Vertical handover enables cellular networks to take advantage of higher bandwidth and lower costs while extending the coverage of cellular networks.

- Network controlled/Mobile station assisted handover
According to the control message service, handover methods can be divided into network control and mobile station assistance. Mobile station assistance is commonly used, the terminal collects channel information and reports it to the BS. The BS decides whether to initiate the handover and which BS should be connected. Then the command is sent to the UE by the BS.

Handover Management in Software-defined Ultra-dense 5G Networks

Hyperdensification is a key method designed to meet the high data traffic of 5G networks. There are a large number of pico BSs in a cell and the coverage radius of a BS becomes smaller than before. Therefore, it is impossible for a single BS to continuously provide high quality services to UE. However, frequent handover is bound to cause a large number of handover delays and higher handover failure rates, which not only affects the quality of UE communications, but also reduces the overall performance of the network system. In addition, since the channel measurement is imperfect, the correct handover timing is difficult to determine.

In [141], a strategy based on SDN-based mobility and available resource estimation was proposed to address the handover delay problem. The specific architecture is shown in Fig. 3.7. The proposed SDN-based ultra-dense 5G network architecture consists of a centralized controller and two independent planes (the control plane and the data plane). In the control plane, the controller can communicate with the mobility management entity (MME) and home subscriber server (HSS) components of LTE to handle the handover procedure. Mobility management module and admission control module are also defined. The mobility management module includes the proposed eNB transition probability estimation engine and an eNB selection engine. A large number of hexagonal small cells and mobile nodes are deployed in the data plane. In addition, two scenarios are considered, the first scenario is that the neighboring six cells of the current cell are in an active state; the second scenario is that

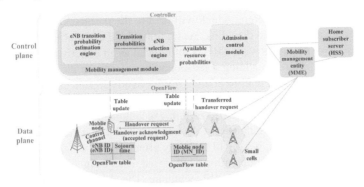

Figure 3.7 The proposed SDN-based ultra-dense 5G network architecture, consists of the centralized controller, and the large number of pico BSs and mobile nodes in two separate planes (control and data planes)

the neighboring six cells of the current cell are in a partially active state, that is, some cells are in a sleep state. Communication between the controller and the data plane is performed with the help of the OpenFlow protocol and the OpenFlow table. Therefore, the decision of the controller is transmitted to the OpenFlow table of the mobile node and the small cell through the OpenFlow protocol. It is determined by the UE residence time whether the current cell UE wants to handover.

In the eNB transition probability estimation engine, a Markov model is used to determine neighboring eNBs with high transition probabilities. In the first scenario, the mobile node on each cell has seven (six neighbor cells and one current cell) different transition probabilities. In the second scenario, the eNB's LTE automatic neighbor relation (ANR) function is used to reach the effective neighbor of the eNB. Each eNB has a neighbor relation table (NRT) and this table is managed by the ANR function. Therefore, the controller can utilize the NRT of the eNB to reach the effective neighbor list of the eNB. Then, as in the first scenario, the transition probabilities of the respective eNBs of these effective neighboring cells are estimated by using a Markov chain. The eNB with the highest transition probability is then transferred to the eNB selection engine.

3.3 SUMMARY

In this section, we discuss the mobility management in UDN. First of all, considering the service cell switching and UE allocation caused by UE mobility, we introduce UE association indicators, such as coverage, spectrum efficiency, energy efficiency, and fairness. In traditional cellular cells, the goal of the UE association scheme is to optimize the relevant indicators. For different schemes, we also introduce three modeling techniques, which are game theory, stochastic geometry, and joint optimization. As joint optimization model can maximize system utility under resource constraints, it becomes the most commonly used method in mobile management. Compared with traditional networks, the transmit power of heterogeneous cells in

UDN is significantly different, which will lead to the problems of unbalanced load and coverage at uplink and downlink. As traditional UE association solutions cannot solve the above problems well, CRE is introduced as an effective mobility management technique. Through adding a reasonable bias value to the SINR or RSRP of the cell signal, UEs will tend to connect to the lightly loaded cells and the load of the macrocell is reduced. CRE is also applicable to multi-layer heterogeneous networks, and is often used with the cell zooming technique. Using high frequency reuse factor as in traditional cellular networks can reduce interference between UEs and increase system capacity, but it will seriously affect QoE at the cell edge. Therefore, the concept of multi-cell cooperation is proposed, such as in 3GPP where multi-cell cooperation is extended to the CoMP technique. The core of CoMP are coordinated scheduling/coordinated beamforming and joint transmission/joint reception. Coordinated scheduling/coordinated beamforming allows CSI sharing between BSs to reduce cell-edge interference. Joint transmission/joint reception not only allows BSs to share CSI, but also allows them to share UE data, which can significantly improve the SINR of UE receiving signals. BSs set selection schemes of CoMP include static, dynamic and semi-static collaboration set partition Semi-static partition combines the advantages in low complexity of static partition and flexible operation of dynamic partition, which has the highest practicality.

Then, we introduce mobility management for uplink and downlink scenarios. In the downlink scenario, switching is the key technique to ensure mobile UEs enjoy consistent QoE. The switching process includes measurement, judgment, and execution. Once the UE cannot obtain better communication service during movement, it needs to switch to other links to ensure the continuity of the communication service. The conditions of handover trigger include: RSS of the serving BS below the threshold, more available bandwidth of target BSs, and higher energy efficiency of target BSs. In addition, handover can be divided into hard handover and soft handover according to the handover mechanism, vertical handover, and horizontal handover according to the network access technology, same frequency handover and different frequency handover according to the switching frequency, network control handover and mobile station assisted handover according to the control difference. In the uplink, the system applies power control to maintain link quality while maximizing total throughput and minimizing interference. We mainly introduce traditional power control, partial power control, open loop power control, and closed loop power control, among which traditional power control and partial power control are divided according to different path loss compensation, open loop power control and closed loop power control are divided according to different ways of obtaining power control information. At present, common power control algorithms include power balance criterion based, SINR balance criterion based, hybrid criterion based, and bit error rate balance criterion based algorithm, among which hybrid criterion based power control algorithm combines the advantages of SINR balance criterion based and power balance criterion based algorithm, thus has the advantage of being highly practical.

Finally, we evaluate the performance of a two-layer heterogeneous network based on uplink CoMP technology, and further understand the decisive role of BS cooperative handover in mobility management.

4 Enabling Factors and Emerging Techniques

4.1 MASSIVE-MIMO

Motivation

Restricted by the limited spectrum resources, subscribers in wireless networks leverage spectrum sharing to make communications more efficient and effective. However, the communication of a significant number of subscribers with the limited spectrum will inevitably lead to interference, even if the frequency reuse scheme is adopted. Moreover, various kinds of fading such as multipath fading and frequency selective fading not only reduce the communication range of subscribers in wireless networks, but also cause problems of inter symbol interference and transmission delay [142]. To address the above challenges, the definition of MIMO is proposed in wireless networks. As a spatial diversity technique, MIMO can significantly improve the capacity and reliability of wireless networks by transmitting multiple streams to obtain multiplexing gain. MIMO technology has been maturely applied in commercial wireless communication systems, e.g., 4G, and it is destined to be integrated into new generation wireless communications standards [143].

Although 5G can use 6 GHz-below frequency bands, limited by bandwidth requirements and low-frequency band resources, most 5G networks have to employ the high-frequency band, i.e., the mmWave band. Compared with centimeter wave and decimeter wave used in 2G/3G/4G, mmWave has more serious signal attenuation. As we know that the received power is proportional to the square of the wavelength. Since the attenuation of mmWave is serious, mmWave will cause an obvious decrement of received power at the antenna. Because the relationship between the antenna size and the wavelength is fixed, the increment in carrier frequency must be at the expense of reducing the size of the antenna. It means more high-frequency antennas must be plugged into the same space. Apparently, the high-frequency path loss can be compensated by increasing the number of high-frequency antennas. The realization of massive-MIMO relies on the above principles. Antenna arrays can generate a high gain and improve network coverage by adjusting beam shape. It is worth noting that the narrow beam can greatly reduce the interference to the periphery.

Due to the advantages in energy efficiency and throughput, massive-MIMO has become a potential competitor of UDN. Meanwhile, UDN and massive-MIMO still have the possibility of cooperation to extend the advantages. The deployment of large-scale antennas at BSs can achieve higher outdoor spatial gain, while the deployment of multiple cells can meet more data transmission requirements. Although massive-MIMO and UDN reflect different network evolution ideas, the combination of these two techniques can achieve large-scale collaboration to greatly improve the network capacity and transmission rate for wireless communications. For example, massive-MIMO can combine numerous cells into a macrocell, then it can transmit data back to the core network. Such transmissions can be regarded as a kind of back-haul for UDN.

Overview of Massive-MIMO

Massive-MIMO technology has received widespread attention after being proposed by Thomas L. Marzetta in 2010 [144]. The basic idea of massive-MIMO is to config-ure large-scale antenna arrays at BSs, in which the number of antennas is hundreds or even thousands. Since the antenna array is several orders of magnitude greater than the existing antenna systems, the large-scale antenna array can serve multiple UEs simultaneously. In fact, only 10% of the total antennas are used to provide service for UEs. Therefore, these antennas can be dispersed into cells or be centralized as an antenna array [145].

Massive-MIMO has multiple physical characteristics and performance advan-tages that traditional MIMO systems cannot surpass. The main benefits are listed as follows:

- Reduced interference
  For Massive-MIMO, each channel is orthogonal to the other channels be-cause of its characteristic of asymptotic orthogonality. Channel orthogo-nalization can effectively eliminate inter-cell interference and improve the total capacity of the network.
- Reduced delay
  Since the Massive-MIMO has the raised number of antennas at the BS, the rapid channel fading and thermal noise can be effectively averaged out. Then, it makes channel less likely to experience deep fading. Consequently, Massive-MIMO can greatly reduce the waiting time of the air interface.
- Improved spatial resolution
  With the deployment of abundant antennas, beam energy can be focused on a small space area, resulting in the improvement of spatial resolution.
- Reduced deployment cost
  Even with the constant envelope signal, massive-MIMO can deploy addi-tional degrees of freedom for beamforming of the transmitted signal. There-fore, the peak-to-average ratio of the transmitted signal can be effectively reduced, so that, the RF front-end can adopt low linearity, low cost, and low power amplifiers. Then the whole expenditure of the system deploy-ment can be effectively reduced.
- Increased array gain
  Applying numerous antennas can improve the gain of the antenna array, thereby reducing the power consumption of the transmitter. Meanwhile, the total energy efficiency of the system can be improved with several orders of magnitude.

Massive-MIMO: A Means of Spatial Densification

Massive-MIMO can be interpreted as a means of spatial densification, in which hundreds or thousands of antennas are equipped at BSs. A large number of excess service antennas is deployed over active terminals in the massive-MIMO system, the extra antennas can achieve time-division duplex operation. Ideally, extra antennas can focus beam energy into a small space to ensure that more UEs can be served within the same resource unit of the given BS. Although massive-MIMO can greatly improve the energy efficiency of communication systems, massive-MIMO still faces serious technical challenges from the perspective of system design and engineering implementation.

- Pilot contamination & channel estimation
  CSI is typically assumed to be obtained by pilot training during uplink transmission and channel reciprocity. However, the limited orthogonal resources force UEs in different cells to reuse the pilots. Then, the estimation of CSI between neighboring cells transmitting pilots at the same frequency will be interfered. The inaccurate estimation on CSI will inevitably result in pilot pollution [146] [147]. Although the number of BS antennas increases to an unprecedented number, resulting in degradation of the effect from additional noise and multipath fading, challenges in UE mobility management and limited available bandwidth have not been effectively solved yet. UE mobility is the main factor affecting the channel consistency time, it can lead to the upper bound of the pilot sequence length. Moreover, due to the limitation of available bandwidth, inter-cell interference caused by pilot pollution will be a key challenge for future communications.
- Inter-cell / Inter-UE interference mitigation
  In order to mitigate inter-cell interference, the traditional interference suppression technique, such as coordinated MIMO, adopts a joint precoding scheme allowing BSs to share global CSI and payload data in real time. However, the number of pilot symbols used for channel estimation is constrained by the channel coherence time. Hence, BSs equipped with large-scale MIMO array are hardly to obtain perfect CSI [148] [149]. Moreover, it is difficult for real-time global CSI at transmitters (CSIT) to implement in practical applications due to the existence of backhaul delay.
- Joint spatial division and multiplexing (JSDM) beamforming
  The accurate acquirement of CSIT depends on explicit downlink training and uplink feedback [150]. However, uplink and downlink transmissions occur in different frequency bands for frequency division duplex (FDD) system. In other words, downlink training and uplink feedback will consume transmission resources in different frequency bands. Note that the signal power in each time-frequency channel at a coherent time slot is proportional to the number of antennas at BS. In order to effectively reduce the length of the channel coherent time slot in the time division duplex (TDD) massive-MIMO system, channel dimensionality reduction techniques must be employed. In literature [151], a scheme named JSDM is proposed to

divide UEs into groups with similar channel covariance feature space. Then, large-scale MIMO gain can be achieved in FDD systems.

- Hybrid beamforming & hardware cost
  Massive-MIMO can achieve high spectral efficiency by deploying M ($M \gg$ 1) antennas at each BS. Meanwhile, M RF chains achieving full gain consume abundant energy. From the perspective of cost-saving, it is possible to deploy M antennas but only S RF chains at each BS, where $S \ll M$. Therefore, a two-stage precoding scheme is proposed, including high-dimensional pure phase RF precoding and low dimensional baseband precoding [152] [153]. As an applicable scheme, it still faces challenges. It is worth discussing the minimum weighted average data rate under the constraints of pure phase RF precoding and the RF chain resources.

- Statistical CSI based beamforming
  A key factor restricting the widespread application of MIMO systems is the increasing complexity of its hardware design and signal processing. Moreover, deployment cost is also an important issue. Considering that each antenna element requires a separate and expensive RF chain during signal receiving and transmitting, an antenna selection scheme, which has attracted considerable attention in the research community, is proposed [154] [155]. In this scheme, a set of available antennas are selected for further processing with fewer RF chains. However, when the selected antennas adopt the diversity technique to obtain full diversity gain, beamforming gain will reduce simultaneously. Aiming to recover most of the beamforming gain by spatial correlation technique, a preprocessing method based on channel statistics is proposed. In general, using channel statistics, the complexity of statistical CSI based beamforming can be effectively reduced.

Wireless Backhaul of UDNs with Massive-MIMO

Reliable and cost-efficient gigahertz backhaul is a prerequisite for achieving the stable connection between macro-cell BS and associated small BSs in UDNs. It has been shown that backhauls with $1 \sim 10$ GHz bandwidth can meet the requirements of UDNs [156]. Traditional optical fiber has the advantages of large bandwidth and high reliability. However, optical backhaul is not an economical choice for operators due to the limitations of deployment and installation. On the contrary, wireless backhaul, especially mmWave backhaul, is more attractive for overcoming geographical constraints. The merit of mmWave is that it has small antenna size and easy signal amplification. Besides, the communication range will be shortened, but the ability of fading resisting will be enhanced when massive-MIMO is deployed in UDN. Therefore, such a combination of mmWave and massive-MIMO in UDN is promising and urgently required, since it is feasible to improve wireless backhaul performance in UDNs. As shown in Fig. 4.1, the mmWave massive-MIMO based backhaul in UDNs can support the communication of numerous small BSs simultaneously by providing multiple streams for each small BS. Although the above combination solution

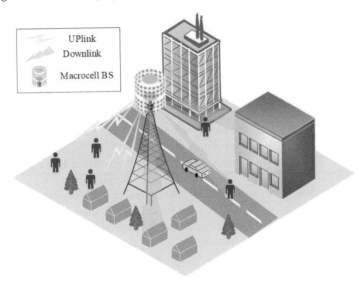

Figure 4.1 MmWave massive-MIMO based wireless backhaul for 5G UDN

improves the performance of the communication system, such a complex scene poses great challenges to the conventional precoding/combining scheme.

The application of the all-digital precoding method in microwave massive-MIMO can support multiple single-antenna UEs simultaneously. It is worth mentioning that each antenna must correspond to an RF link. The raised number of antennas will inevitably lead to an increase in system complexity and implementation costs, resulting in hindering the application of the massive-MIMO technique. Traditional massive-MIMO adopts the analog precoding scheme, which uses a low-cost phase shifter to control the phase of each antenna transmit signal. However, this scheme is only applied in the single-user MIMO system with a single stream. Recently, a hybrid scheme consisting of analog and digital precoding/combining has been proposed, which can effectively reduce transceiver cost and complexity. However, this scheme is typically appropriate for the single-user MIMO system with multiple streams or the multi-user MIMO system with a single stream.

Gao *et al.* [157] propose a hybrid precoding/combining scheme for digital phase shifter network (DPSN). This scheme can be applied in mmWave massive-MIMO based wireless backhaul to support multiple UEs and multiple streams in UDN. Specifically, it is assumed that the macrocell BS has N antennas and each small BS has M antennas. The number of small BSs simultaneously supported in the network is denoted by K. In addition, the mmWave massive-MIMO channel matrix between macrocell BS and the K-th small BS is denoted by $\mathbf{H_k}$ ($\mathbf{H_k} \in \mathbb{C}^{N \times M}$, where $N > M$). Furthermore, the channel matrix $\mathbf{H_k}$ can be expressed in terms of singular value decomposition (SVD), which is given as follows,

$$\mathbf{H_k} = [\mathbf{U_k^1}|\mathbf{U_k^2}] \begin{bmatrix} \boldsymbol{\Sigma_k^1} & \mathbf{0} \\ \mathbf{0} & \boldsymbol{\Sigma_k^2} \\ \mathbf{0} & \mathbf{0} \end{bmatrix} \begin{bmatrix} (\mathbf{V_k^1})^* \\ (\mathbf{V_k^2})^* \end{bmatrix} \approx \mathbf{U_k^1} \overset{1}{\underset{k}{\sum}} (\mathbf{V_k^1})^* \qquad (4.1)$$

The precoding matrix $\mathbf{P}_k = (\mathbf{U}_k^1)^*$ and combining matrix $\mathbf{C}_k = \mathbf{V}_k^1$ can be used to efficiently implement independent multi-stream transmissions. Considering the processes of precoding, digital precoding matrix $\mathbf{P}_{d,k} \in \mathbb{C}^{R_k \times R_k}$ and analog precoding matrix $\mathbf{P}_{a,k} \in \mathbb{C}^{R_k \times N}$ can be approximately cascade to $\mathbf{P}_k$, where $\mathbf{P}_{d,k}$ and $\mathbf{P}_{a,k}$ can be obtained by iterative methods. Similarly, digital combining matrix $\mathbf{C}_{d,k}$ and analog combining matrix $\mathbf{C}_{a,k}$ can be obtained by the same method. Moreover, the downlink channel matrix of multi-user MIMO between macrocell BS and K small BSs denoted by $\mathbf{H}$ ($\mathbf{H} \in \mathbb{C}^{N \times M}$) can be further expressed as $\mathbf{H} = [\mathbf{H}_1, \mathbf{H}_2, ... \mathbf{H}_k]$, where $\mathbf{H_k} \approx \mathbf{U}_k^1 \boldsymbol{\Sigma}_k^1 (\mathbf{V}_k^1)^*$ for $1 \leq k < K$. Furthermore, the analog precoding matrix at the macrocell BS is denoted by $\mathbf{P}_a$, where $\mathbf{P}_a = [\mathbf{P}_{a,1}^T|\mathbf{P}_{a,2}^T|...\mathbf{P}_{a,K}^T]^* \in \pounds^{K_R \times N}$.

Then, the proposed precoding/combining scheme can diagonalize the equivalent channel $\mathbf{P}_d\mathbf{P}_a\mathbf{H}$ diag$\{\mathbf{C}_1, \mathbf{C}_2, ..., \mathbf{C}_K\}$, where $\mathbf{C}_k = \mathbf{C}_{a,k}\mathbf{C}_{d,k}$. This scheme is different from existed solutions, since it can realize transmissions for multiple UEs with multiple streams. Moreover, as the mmWave massive-MIMO channel matrix has obvious low-rank property, the performance losses of the proposed precoding/combining scheme with the reduced number of baseband chains is negligible.

4.2   MACHINE LEARNING

Motivation for Machine Learning

One of the dominating features of 5G network is high flexibility. The overall network system can be separated according to different applications and business requirements. These logically independent networks have specific characteristics and functions. To realize the intelligence of 5G, all layers of the network need automatic resources allocation, adjustment, optimization, and restoration. With the deployment of cloud computing platforms on 5G, there is no doubt that such functions can be achieved in the network.

As a promising technique, machine learning is considered to be an effective enabler to achieve functions such as network slice management, dynamic resource allocation, and self-located fault repair in 5G network. Based on cloud computing platforms, machine learning plays a key role in the achievement of the 5G vision. For example, machine learning enables 5G networks to have the ability to achieve real-time applications such as remote surgery and wireless virtual reality. In fact, machine learning is driven by big data. The comprehensive knowledge must be acquired from big data through machine learning [158]. Therefore, the level and quality of big data are directly and closely related to the development of machine learning. Essentially, the progress of big data is inseparable from the support of 5G networks which can provide large-scale data stream transmission with high speed and low latency. As a result, machine learning, 5G communication, and high-performance computing are mutually reinforcing.

Machine Learning

Machine learning is a field using statistical techniques to give computer systems the "learning" ability from data without being explicitly programmed, e.g., progressively improving performance on a specific task. The term "machine learning" was coined in 1959 by American pioneer Arthur Samuel [159]. Machine learning is regarded as noteworthy progress for the exploration of AI. However, due to the increasing emphasis on logical and knowledge-based methods, machine learning has been reorganized into a separate field, which aims at solving practical problems.

Machine learning is closely related to optimization, since the goal of most learning tasks is to achieve the minimization of the loss function in the training set. The difference between machine learning and optimization is the overall goal. Optimization algorithms can minimize the loss of training sets, while machine learning focuses on minimizing the loss of invisible samples [160]. Actually, the method used by machine learning is similar to that used by data mining. However, machine learning has advantages in prediction by the known attributes learned from the training set, while data mining concentrates on discovering unknown attributes in the training set.

Machine Learning Tasks and Applications in UDNs

Machine learning can be simply categorized as supervised learning and unsupervised learning. Herein, "supervised/unsupervised" indicates whether tagged samples exist in the dataset [161]. Note that supervised/unsupervised learning can complete most learning tasks for machine learning. However, these two learning models are quite different from the human learning process which is to proactively test the environment. The human learning process is named reinforcement learning. Particularly, reinforcement learning is a learning process with continuous adjustment, where agents make evaluations and summaries according to the environment feedback to improve policy decisions. Then, the new environment feedback will be generated according to the new behavior. Next, the basic concepts, models, and corresponding applications of machine learning tasks are described below based on the categories of supervised learning, unsupervised learning and reinforcement learning. Furthermore, the basic features and applications of learning algorithms are summarized in Table 4.1.

- Supervised learning
  Supervised learning is a machine learning mechanism that learns mapping function from inputs to outputs based on example input-output pairs. The prediction model in the supervised learning needs continuous adjustment, and the prediction results are used to compare with the labels until the accuracy of the model reaches the threshold. In general, supervised learning can effectively solve classification problems and regression problems. The common supervised learning method are introduced as follows.

Table 4.1

Machine Learning Algorithms and Applications

Category	Learning Techniques	Applications
Supervised learning	Regression models	Energy learning [162]
	K-nearest neighbor	Handover [161], Energy learning [162], Cache strategy [170]
	SVM	Handover [161], MIMO channel learning [171]
	Bayesian learning	Massive MIMO learning [163] Cognitive spectrum learning [164, 172, 173]
Unsupervised learning	K-means clustering	CoMP [165], Cache strategy [170]
	PCA	Fault detection [161], Smart grid [166]
	ICA	Fault detection [161], Spectrum learning [174]
Reinforcement learning	MDP	EH [167]
	Q-learning	Interference management [168, 175]
	Multi-armed bandit	Resource allocation [169]

- Regression models

  Regression models such as K-nearest neighbor (KNN) and support vector machine (SVM), can be adopted for the prediction of radio parameters associated with particular UE. For example, the above models are applicable to solve high-dimensional search problems in detection and channel estimation in massive-MIMO systems. Moreover, SVM model can estimate channel noise statistics with training data. Furthermore, both KNN model and SVM model can provide the best handover solution for UDNs which have frequent handover tasks. According to the experiments in [162], KNN algorithm can efficiently achieve energy demand prediction with 90% accuracy.

- Bayesian learning

  Bayesian learning model can be used for spectral feature learning and estimation in UDNs. In order to solve the pilot pollution problem encountered in massive-MIMO systems, the authors of [163] estimate the channel parameters of the target cell link and the interference links with sparse Bayesian learning techniques. Similarly, the authors of [164] propose a cooperative broadband spectrum sensing scheme, which can be adopted to detect the primary UE supported by the multi-antenna assisted cognitive radio network.

- Unsupervised learning

  Unsupervised learning is a machine learning mechanism that learns from the data without label, classification or categorization. Unsupervised learning can effectively characterize the structure and patterns in the input data. Thus, it is widely adopted in scenarios including association rules learning and clustering.

- K-means clustering
  Clustering is a common method in UDNs. For example, for the sake of avoiding mutual interference, it is necessary to apply a coordinated multipoint transmission scheme to cluster the cells. Since the best unload strategy is adopted among mobile UE clusters, device clusters can achieve high energy efficiency in D2D networks. Harai *et al.* [165] propose a hybrid optical/wireless network scheme, which reduces long-distance traffic by encouraging the use of high-capacity optical infrastructure. In order to solve the mixed integer programming problem in system optimization, the classical K-means clustering is applied to divide grid access points into several groups, thereby jointly improving the network partition and virtual channel allocation.
- Principal component analysis (PCA) and independent component analysis (ICA)
  Both PCA and ICA belong to powerful statistical signal processing techniques, aiming at restoring independent source signals statistically from linear mixtures. The main applications of PCA and ICA in the field of UDNs include anomaly detection, fault detection, and intrusion detection, all of which rely on traffic monitoring. In addition to the above applications, PCA and ICA can also be deployed in physical-layer signal size reduction for massive-MIMO systems. In the smart grid scenario, Qiu *et al.* [166] adopt PCA and ICA to recover the simultaneous wireless transmissions of smart utility meters deployed in every family. Before decoding, signals received from all smart meters must complete the separation procedure at the power utility station. ICA can be used for blind separation of the signal statistical properties. Specifically, it can enhance transmission efficiency by avoiding channel estimation in each frame while improving data security by eliminating interference in wideband and signals.
- Reinforcement learning
  Reinforcement learning is a machine learning mechanism concerned with how software agents ought to take actions in an environment so as to maximize the cumulative reward. Reinforcement learning makes it possible to find the best decision among a large number of possible paths, the common application scenarios of which include dynamic systems and robot control.
- MDP
  MDP is an ideal tool to support UDN decision making, where network constitutes the environment and UEs can be regarded as agents. The classic applications of MDP in UDNs include network selection, association issues and EH. Most of these application scenarios are under the condition of time-varying channels and limited battery. In order to achieve resource optimization in such complex environments, battery utilization or channel selection is usually set as the action. In literature [167], Aprem *et al.* adopt MDP to solve the transmission power control problem in EH system, where the state space is defined by battery state, channel state, and packet transmission/reception state. Because the CSI can only be observed partially, the

machine learning policy and the voting heuristic policy are proposed to obtain suboptimal solutions. Based on the belief state of the channel and the partially observable MDP model, the communication quality is guaranteed, but the power consumption is reduced.

- Q-learning
  Q-learning is usually used in combination with MDP. A classic application example in UDNs is cell interference management and compensation [168]. In order to reduce interference through the Q-learning method, the state, action, and reward function of the system must be related to the channel state and transmission power. Specifically, the system state is determined by channel quality and resource blocks allocated by small cells. The system action is used to control downlink power. Moreover, the policy consisting of action decision-making can be effectively adjusted according to the reward function. Utilizing the model trained by Q-learning to achieve interference management in the communication system, it can be found that the proposed strategy achieves better performance.
- Multi-Armed Bandits (MAB)
  As an emerging signal processing tool, MAB can solve the challenge in resource allocation. The unknown channels or other wireless environment parameters must be "explored" continuously, while the known channels must be "exploited" by UEs in the group. Generally, MAB model is applicable in multi-player adaptive decision-making problems, in which the best joint action model can be obtained by the continuous interaction between players and the dynamic environment, thereby achieving an equilibrium eventually [169].

In addition to the above types, with the continuous development of machine learning, some hybrid applications have also appeared in this field. For example, the learning model trained by a part of the labeled data is called semi-supervised learning. Besides, the model classification of machine learning is more plentiful. A typical example is neural network, it is a mathematical model for information processing, whose structure is similar to the synaptic connections in brain. The neural network was considered to be a kind of supervised learning model when it was first proposed. However, with the evolution of deep learning, a variety of neural network algorithms have emerged and some of them even involved unsupervised learning.

Machine Learning Based Small Cell Cache Strategy for UDNs

In order to improve spectral efficiency in UDNs, the macrocell offloads mobile data traffic to dense small cells. However, it will inevitably pose challenges to network backhaul links between small BSs and core networks. Moreover, due to the limited radio spectrum resources, wireless backhaul must endure greater pressure from substantial amounts of data traffic. In fact, the contents requested by UEs are usually duplicates. Therefore, the proactive caching technique achieving predictive storage based on popular content in BSs, can satisfy the demand of UEs in overhead and content-providing. Since proactive caching has an excellent performance in backhaul congestion mitigation and QoE advancement, it has become a popular method in cache strategy.

With the gradual maturity of the proactive caching technique, most studies focus on performance improvement by traditional methods, e.g., optimization algorithms and stochastic geometry caching. However, the traditional solutions often work with ideal assumptions but do not work well in the real scenarios. Moreover, it is difficult for these solutions to achieve significant performance breakthroughs. In order to seek innovative and practical schemes to solve the above issues, Shen *et al.* [170] consider the joint application of machine learning including KNN and K-means. K-means is a clustering algorithm in unsupervised learning, while KNN is a classification algorithm in supervised learning. In fact, K-means is the simplest but most efficient clustering algorithm, which is similar to KNN in label classification based on neighbor information. The core idea of K-means is that k initial centroids need to be specified as the clusters of clustering, then the iterative process is executed repeatedly until the algorithm converges. However, for the KNN method, the distance between the feature values of the new data and those of the training data needs to be calculated. Then, $K(K \geq 1)$ nearest neighbors for classification or regression are selected. Note that, if $K = 1$, the new data will be assigned the same category as its neighbor. Different from the optimization methods based on mathematical models, solutions with machine learning are driven by real-world data, therefore they are less sensitive to model defects.

Figure 4.2 shows a wireless communication model considering two-tier UDNs. We denote UE set, macrocell set, and small cell set as $\mathbf{N}$, $\mathbf{M}$ and $\mathbf{P}$, respectively. In particular, the number of elements in $\mathbf{N}$, $\mathbf{M}$ and $\mathbf{P}$ are N, M and P, respectively. The wireless backhaul capacity of the small BS $p(p \in \mathbf{P})$ is denoted by B_p. The number of UEs associated with small BS p is denoted by N_p. The content set requested by all UEs is represented as $\mathbf{C}$. Therefore, the size of the contents in $\mathbf{C}$ can be expressed as $\mathbf{s} = [s_1, s_2, ..., s_C]$, where C represents the total number of contents, calculated by $C = |\mathbf{C}|$. The cache policy matrix at time t is expressed as $\mathbf{X}(t) \in \mathbb{R}^{P \times C}$, where $[\mathbf{X}(t)]_{pc} \in \{0, 1\}$. It is worth mentioning that $[\mathbf{X}(t)]_{pc} = 1$ indicates that the p-th small

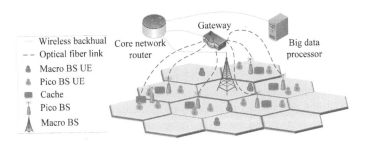

Figure 4.2 Machine learning based small cell cache strategy scenes

BS caches the c-th content. Then, the objective problem can be expressed as

$$\min_{\mathbf{X}(t)} \quad \Phi(\mathbf{X}(t))$$

$$s.t \quad \Sigma_{c=1}^{C}[\mathbf{s} \odot \mathbf{x}_p(t)]_c \leq \mathbf{S}, p \in \mathbf{P}, \tag{4.2}$$

$$[\mathbf{X}(t)]_{pc} \in \{0,1\}, p \in \mathbf{P}, c \in \mathbf{C}$$

where $\mathbf{x}_p(t)$ represents the p-th row of $\mathbf{X}(t)$. The object function $\Phi(\mathbf{X}(t)) = \frac{\Sigma_{c=1}^{C}[\mathbf{s} \odot \mathbf{f}]_c}{\Sigma_{c=1}^{C}[\mathbf{s} \odot \mathbf{g}]_c}$ is the system backhaul load where $[\mathbf{f}]_c = \Sigma_{p=1}^{P}[\mathbf{A}_l(t,\tau) \odot \mathbf{1} - \mathbf{X}(t)]_{pc}$, $[\mathbf{g}]_c = \Sigma_{p=1}^{P}[\mathbf{A}_l(t,\tau)]_{pc}$, $\odot$ denotes *Hadamard product* and $\mathbf{A}_l(t,\tau)$ denotes the content request times matrix during the time interval $(t - \tau, t]$ in the l-th day.

The cache efficiency optimization problem can be further reformulated into a system backhaul load minimization problem. However, it is difficult to solve such a problem under the condition of highly random cache content. Hence a machine learning-based cache strategy is proposed to solve the problem from the perspective of exploiting the potential of mobile traffic data. Firstly, the K-means clustering algorithm is used to fully reveal the hidden spatiotemporal patterns of small BSs in content requestion. Therefore, both the personalized caching between clusters and the predictive caching within a certain cluster can be satisfied. Secondly, the KNN classification algorithm is introduced to categorize contents. Therefore, the new contents can be periodically cached in the corresponding cluster with high precision and low complexity. The analysis of simulations indicates the proposed caching strategy is better than the existing methods in performance improvement.

Continuous Time Markov Chain Based Reliability Analysis for UDNs

In 2G, 3G, and 4G communication systems, configuration and maintenance parameters required for cells are about 500, 1000, and 1500, respectively. It can be inferred that the cells in 5G will maintain the same growth trend, which will pose great challenges in network management. SON is regarded as an inevitable measure for achieving effective and reliable management of the complex networks. However, when multiple SONs work simultaneously, they may suffer from potential conflicts caused by parameter overlap and objective coupling, thereby affecting cell outage rate in UDNs. Moreover, factors such as hardware/software failures and supplier incompatibility, also pose significant challenges for the reliable operation in UDNs.

Farooq *et al.* propose a stochastic analytical model to analyze the effects of faults arrival in cellular networks [176]. They exploit continuous time Markov chain (CTMC) with exponential distribution to model BS reliability behavior in failures and recovery times. After subsequent analysis of the model, an adaptive failure prediction framework is proposed. Since the framework learns from the previous failures database dynamically, it can effectively reduce network recovery time and improve system reliability.

The occurrence and recovery of faults usually result in the transition from one system state to another. Since the exponential random variable is the only continuous variable with Markov property, it can be assumed that the time of transition follows the exponential distribution. The term $X(t)$ with finite state space $\mathbf{S} = \{1,2,3\}$ represents the state of a BS at time t.

- When $X(t) = 1$, it indicates that the BS is in a healthy state at time t. Meanwhile, all parameters are configured with the optimal values
- When $X(t) = 2$, it indicates that the BS is in a suboptimal state at time t, where at least one parameter is misconfigured. In such a state, the performance will drop below the typical level if the BS continues to work
- When $X(t) = 3$, it indicates a complete interruption at time t

It is assumed that the time of failure follows the exponential distribution and the arrival rate of faults is temporarily independent. When exploiting Poisson distribution for model design, failures can be classified as trivial failures characterized by arrival rate λ_t and critical failures characterized by arrival rate λ_c. Trivial failures will drive the network from an optimal state to a suboptimal state, while critical failures will cause issues of complete interruption. In a bad state, the BS can use a self-coordinating framework-driven recovery module to achieve autonomous adjustment. Specifically, the recovery from the suboptimal state to the optimal state must go through the process of cell anomaly detection, cell diagnosis, and compensation. However, the recovery from the complete outage only involves the compensation time in most instances. Therefore, the recovery time from the suboptimal state follows the exponential distribution with mean value $1/\mu_{dc}$, and the recovery time from a complete outage state follows the exponential distribution with mean value $1/\mu_c$. Since the transition between failure and recovery is only determined by the current state instead of the path to the current state, the transient process $X(t)$ can be mathematically modeled as a temporally uniform CTMC over the state space $\mathbf{S}$. For each time t, the probability of the BS in state j is given by:

$$pr_j(t) = Pr\{X(t) = j\}, j \in \mathbf{S} \tag{4.3}$$

The row vector of the transient probability of $X(t)$ is denoted by pr_t, where $pr_t = [pr_1(t), pr_2(t), pr_3(t)]$. Moreover, the generator matrix $\mathbf{Q}$ and the rate matrix $\mathbf{R}$ of $X(t)$ can be expressed as follows, respectively,

$$\mathbf{Q} = \begin{bmatrix} -\lambda_t - \lambda_c & \lambda_t & \lambda_c \\ \mu_{dc} & -\mu_{dc} - \lambda_c & \lambda_c \\ \mu_c & \mu_c & \mu_c \end{bmatrix}, \tag{4.4}$$

$$\mathbf{R} = \begin{bmatrix} 0 & \lambda_t & \lambda_c \\ \mu_{dc} & 0 & \lambda_c \\ \mu_c & 0 & 0 \end{bmatrix} \tag{4.5}$$

Based on the established model, performance indicators used to quantify network reliability including occupancy time, first passage time, and steady-state distribution can be calculated. Relying on these indicators, the fault prediction framework realizes efficient construction, as shown in Fig 4.3. The framework estimates λ_t, λ_c, μ_{dc},

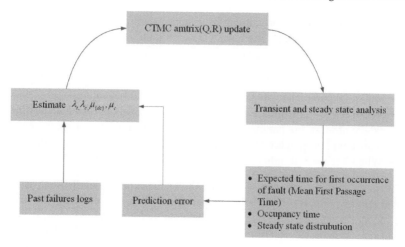

Figure 4.3   Fault prediction framework

and μ_c by machine learning tools from historical parameter data of BSs in the cellular network. Hence, the mean values can be introduced into the CTMC model. Then, the generator matrix **Q** and the rate matrix **R** can achieve dynamic updates.

The proposed framework can minimize the degradation time by accurately estimating the suboptimal probability of a cell within the given time. Once the predicted failure time approaches, priority verification for each BS configuration parameter will begin. Moreover, the steady-state distribution of the elapsed period can be used as a KPI for cell performance. If the time that a cell being the suboptimal state or the interruption state is higher than a threshold during the optimization process, the corresponding prioritization sort should be executed. Furthermore, the proposed framework can also be used in failure diagnosis. If records can be partially retained between the occurrence of the failure and the corresponding source of the failure, the diagnosis can begin with the root which has been recorded in the table. The above diagnosis process occurs before the advent of the expected suboptimal state or interruption state. Therefore, the time in diagnosis and compensation can be greatly shortened.

4.3   SOFTWARE DEFINED NETWORK (SDN)

The Goal of SDN

UDNs are expected to be dynamic, cost-efficient, and flexible. In the existing networks, the traffic control and forwarding, the operating system and dedicated hardware coupled with the service characteristics are integrated on the network devices. However, the aforementioned operating system and dedicated hardware are developed and designed by different manufacturers, resulting in inevitable severe coupling issues. As we can imagine, with the increasing number of terminal types and service types, the scalability of this "big behemoth" in function and service will be difficult to be effectively improved. Therefore, such traditional network architectures indeed pose challenges in both cost and time for operators to deploy network services.

To meet various network requirements, communication systems must strip the physical topology of the network, virtualize network resources, and make the management in logical layer uniform [177]. These operations aim to get rid of the hardware limitations on the network architecture, which is in line with the core idea of SDN. In SDN, applications can be efficiently installed, and the network architecture can be conveniently modified like software updating. Presuming that the existing network is a mobile phone. Then, the goal of SDN is to design an operating system (e.g., Android) to provide powerful services for the "network world" by various mobile Apps.

The Basic Feature of SDN

In 2012, the Open Network Foundation released the white paper *"Software Defined Networking: The New Norm for Networks"*. In the white paper, the following definitions are given to SDN: *"SDN is an emerging network architecture with separate control, forwarding and direct programming. The core is to decouple the traditional network architecture with tightly coupled devices into a three-layer separation architecture of application, control, and forwarding. At the same time, centralized control of network and program ability of network applications need to be achieved through standardization"* [178]. Compared with traditional network architectures, the main technical features of SDN are reflected in three aspects [179]:

- Separated forwarding and control
  Forwarding and control are not integrated into SDN. Specifically, the SDN controller is designed to realize functions, such as network topology collection, routing calculation, traffic table generation, and network management, while network layer devices are only responsible for traffic forwarding and policy enforcement. Therefore, the planes of forwarding and control can be developed independently in the network system. The forwarding plane needs to be optimized towards generalization, simplification, and low cost, while the control plane needs more capacity to support the development of centralization and unification.
  The separation of SDN control and forwarding greatly reduces hardware costs, thereby promoting the application of SDN. However, since frequent changes in device hardware and forwarding flow tables bring device compatibility issues between SDN and existing networks, the application of SDN in large networks may be restricted within a certain period.
- Centralized control logic
  Since the global static topology, dynamic forwarding tables, resource utilization information, and fault status of the entire network can be easily obtained in the centralized control plane, SDN controller can implement unified management, control, and optimization at the network level. Moreover, the SDN controller can further implement fault location and elimination

according to the dynamic forwarding information of the global topology, thereby improving operational efficiency.

Centralized SDN control can achieve global optimization and provide end-to-end network deployment, assurance, and detection. In addition, the SDN controller can centrally manage the network with distinct levels to achieve the coordination and optimization of the multi-layer and multi-domain network. A classic example is the joint scheduling of packet networks and optical networks.

- Open network interface
  An important feature of SDN is to support the open network interface. Resources in SDN can be centrally managed, integrated, and virtualized with the centralized SDN controller, while network services can be provided by a standardized northbound interface to further improve the network capabilities.

  The open network interface in SDN simplifies the programmability of the network and makes application services easier to provide. Herein, the network is not only a kind of infrastructures, but also a service provider to further expand the scope of SDN applications.

The separation of forwarding and control can effectively reduce the hardware cost of the device. The centralized control logic can achieve system optimization, multi-network convergence, and centralized management with a global perspective. The open network interface can encourage more innovation in business and network service. These three driving forces have promoted the development of SDN and enabled SDN to have more application scenarios as well.

CROWD: An Energy-Efficient SDN Approach for UDNs

The densification of UDNs is hard to achieve only by rescaling the existing protocols/networks due to the following reasons: (1) Wired backhaul have difficulties in scalability. Installing a new wired infrastructure requires more investment than deploying a wireless access point (AP). However, wireless backhaul is not compatible with wired backhaul; (2) The existing protocols are only suitable for efficient operation under the current network density. System performance may decrease as the scale of the network further increases; (3) In terms of energy consumption, large-scale deployment of BSs and interconnected network components brings an extra burden on operators and the whole communication system.

Undoubtedly, SDN is the best solution to solve the above problems. Specifically, SDN not only has a professional and fine-grained optimization mechanism, but also provides dynamic resource allocation. Ali-Ahmad et al. [150] adopt a dynamic two-layer SDN controller to achieve connectivity management. In the connectivity management for energy optimised wireless dense networks (CROWD), the CROWD local controller (CLC) can quickly make decisions within a limited and fine-grained range, while the CROWD regional controller (CRC) can slowly make decisions within a wider but rougher range. Moreover, these two management methods can ensure the aggregation of control information without providing an extensive range of detailed information, thereby reducing the signal overhead.

The proposed network architecture encompasses LTE (macro/pico/femto) and WiFi cellular, and the architecture is structured into two logical tiers by CLC and CRC. Each district consists of interconnecting backhaul links and BSs. Actually, the BSs are LTE eNBs and WiFi APs. The application programming interfaces (APIs) that achieve connection with the CLC are called NB interfaces, which can optimize the operations within a district. Specifically, NB APIs have at least the following two categories: technology-specific NB API and technology-agnostic NB API. Technology-specific NB API exposes fine-grained details obtained from BSs and offers methods which are only valid for the specific communication protocol. Technology-agnostic NB API exposes abstraction and aggregated data and offers generic modifiers which may be valid for a wide range. Any application can connect to one or more APIs depending on its optimization goals and requirements. Meanwhile, the CLC can access different southbound interfaces of BSs to control wireless operation and the backhaul network. The SB is composed of three interfaces has the following functions in the CRC, respectively, backhaul network management, CLC controlling in the region and information exchange. Figure 4.4 shows the architecture of CROWD, in which the NB interfaces and the SB interfaces among the CLC, BSs, and backhaul are described clearly. Moreover, the key interconnections between new and existing network elements have been highlighted as well. For example, the eNBs have split connections in the case of LTE, where the control path (via the 3GPP S1-MME and X2 interfaces) goes through the CLC entirely while the data path is directed to the distributed mobility management gateway. Split connection is a novel feature in CROWD, based on which the CLC application can interject the communication between the eNB and the MME, thereby anticipating or overriding

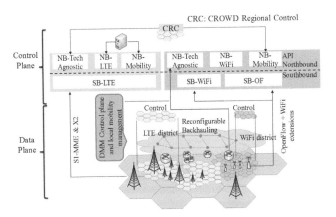

Figure 4.4 CROWD network architecture

the mobility decisions in the district. Although the standard assumes that information should be exchanged between peer eNBs, the X2 interface has already supported the communication with a wide range of data. Therefore, it is proposed to use X2 interfaces to collect fast and detailed measurements from the LTE eNBs.

The main building blocks in the above architecture include access selection and the splitting between control planes and data planes. The separated control path and data path in SDN achieve effective control of actual data stream essentially and optimize the connection management in UDNs. In order to discover greater advantages of SDN, increased studies begin to focus on the schemes in data path control through different networks.

WiSEED: Another SDN Approach for UDNs

Wireless software-based architecture for extremely dense networks (WiSEED) [180] is another management architecture with the support of SDN in UDNs. The unique framework is designed to meet key requirements in scalability, flexibility, and energy efficiency for future dense networks. Specifically, modules in WiSEED co-manage a set of operational services including routing, mobility, and spectrum usage, all of which can be regarded as software running on an SDN controller.

Figure 4.5 describes the architecture of WiSEED, including planes of energy efficiency, scalability, and resilience. These planes represent the requirements of the 5G networks. Meanwhile, the main operational services in WiSEED include routing, mobility, and spectrum management. The operational services combined with mechanisms, algorithms, and technologies achieve better cooperation to make a reasonable tradeoff among the planes, thereby realizing ubiquitous and seamless access among mobile UEs effectively. Moreover, Fig. 4.5 also shows the actions of routing

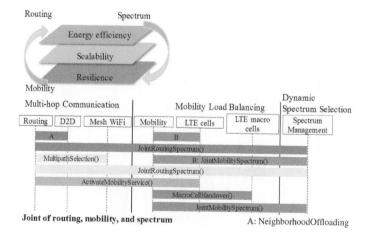

Figure 4.5 WiSEED – architectural model, joint of routing, mobility, and spectrum

and mobility in detail. For example, the sequential distribution of traffic load and mobile devices from a micro-level (e.g., D2D) up to a macro-level (e.g., LTE eNB), and the interaction with dynamic spectrum management. It is worth noting that the actions mentioned above are managed by WiSEED software modules in the SDN controller.

In order to optimize the routes with the light load and the best signal quality, WiSEED first adopts the routing operation service based on multi-hop and direct communication. Generally, an initialized session in the local neighborhood is circulated by D2D. The routing service needs to open a more comprehensive view through Mesh WiFi communications, regardless of whether it participates in the demand plane. If the above routing scheme fails to meet the requirements, the mobility operation service will perform handover initialized by small cells.

Spectrum management services realize dynamic spectrum allocation based on demand plane and spectrum availability. For example, one channel may be repeatedly allocated between close links in dense scenarios. However, the channel sharing between licensed frequency bands and unlicensed bands leads to insufficiency of non-overlapping channels and increases the competition. Since the routing service only makes partial advantage of network density to unload traffic, the traffic load balancing is hard to reach the desired effect. Therefore, spectrum management services should work together with routing and mobility services in cross-layer optimization to provide additional available channels for cognitive devices through spectrum databases.

Moreover, spectrum management can provide quick link connection with devices to change communication characteristics, such as channel bandwidth, transmission power, and beamforming. In the emerging spectrum sharing system, PUs can rent or auction free channels to SUs through the auction process. The combination of auction mechanisms and dynamic spectrum management services can meet more requirements for UEs. If the spectrum of a particular link or cell antenna is not available in the spectrum database, spectrum management services will activate an auction mechanism for purchasing spectrum. Similarly, if an idle spectrum is licensed and the mobile operator network has the option to rent or auction the spectrum, spectrum management services will activate the auction mechanism for spectrum sales.

4.4   CLOUD-RADIO ACCESS NETWORK (C-RAN)

The background of C-RAN

A radio access network (RAN) is part of a mobile telecommunication system to connect UEs and the core network. RAN can provide 24×7 hours of uninterrupted, high-quality services to UEs. Therefore, RAN is determined as an important asset for the survival of mobile operators. The traditional RAN has the following characteristics: (1) BSs connecting with fixed numbers of sector antennas only have a small coverage area. It means that the service range of sending and receiving signals is limited to the local area; (2) Independent BSs cause low spectrum efficiency, since each BS can only work with its own spectrum. Therefore, the limited spectrum makes increased co-channel interference, resulting in reduced channel capacity; (3) BSs are usually developed on proprietary platforms.

The above three characteristics bring challenges to the deployment of the system in cost and performance. Firstly, the dense deployment of BSs brings prohibitive costs in construction, site leasing, maintenance, and operation. Moreover, the average load of the network is always much lower than the busy hour load. Therefore, the whole utilization of each BS is still low. Finally, mobile operators need to maintain multiple incompatible platforms. It means higher costs are needed for platform capacity expansion or upgrading. To meet the growing demand for mobile data services, mobile operators need to constantly upgrade the networks with multiple standards. However, the network upgrading on proprietary platforms has poor flexibility, hence, it is difficult to efficiently upgrade the networks in practice.

Actually, the construction and operating costs have brought enormous pressure on revenue for operators. The fierce competition in the telecommunications market has slowed down or even declined the growth of average revenue per UE [181]. Meanwhile, the overall development of the entire industry has been affected by the compression of investment in network construction and equipment procurement. Under such conditions, mobile operators must find a solution to provide UEs with low-cost and high-performance wireless services. Then, the communication industry can be further promoted on sustainable profitability and long-term development.

In April 2010, China Mobile released the C-RAN white paper on the new wireless network architecture for green evolution officially, expounding the future development vision of centralized baseband processing network architecture technology. C-RAN refers to the green wireless access network architecture based on centralized processing, collaboration radio, and real-time cloud Infrastructure [182]. Different from traditional distributed BSs, C-RAN breaks the fixed connection between remote RF unit (RRU) and baseband unit (BBU). RRUs are deployed near the antennas to greatly reduce attenuation caused by the feeder. Meanwhile, BBUs are migrated to the central machine room uniformly to form a BBU pool. Then, RRUs can connect with the central machine room through the pre-transmission network and create favorable conditions for inter-cell cooperation.

As shown in Fig. 4.6, the C-RAN architecture consists of three main components:

- Distributed network composed of RRUs and antennas
- High broadband and low latency optical transmission networks connecting RRUs and BBUs
- Centralized baseband processing pool composed of high-performance processor and real-time virtual technology

Based on the above three components, the performance of the network can be further improved. Firstly, the distributed RRUs can support wide coverage of large-capacity wireless networks due to the advantages of light weight, low cost, and easy installation [183]. Secondly, high-broadband and low-latency optical transmission networks achieve the effective connection of BBUs and RRUs. Thirdly, the baseband processing pool has powerful processing capabilities because of the aggregation of high-performance processors and real-time virtual technology. Moreover, centralized

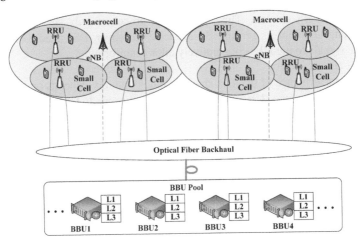

Figure 4.6 C-RAN architecture

deployment of baseband processing greatly reduces the requirements for the machine room and enables the implementation of resource aggregation and large-scale collaborative wireless transceiver technology.

C-RAN: A Centralized Dense Network

Since the dense deployment of RRUs, C-RAN is regarded as a centralized dense network, which is different from the naturally distributed dense small cell network. Although C-RAN is generally considered as a competitor to the dense small cell network, these two kinds of networks can still be coexisted. For example, deploying both C-RAN and the dense small cell network in homes or office buildings without fiber resources optimizes the densification of cells. Due to the advantages of system cost, capacity, and flexibility, compared with traditional wireless access networks, C-RAN can be applied in crowded areas such as shopping centers, concerts, and stadiums, to provide better experiences for UEs [184]. However, the development of C-RAN still faces the following challenges in the industry:

- Backhaul capacity
  As all BSs are connected to the central processor through a digital backhaul link in C-RAN, joint data processing and data precoding across multiple BSs can be implemented. However, the successful execution of related functions requires the sharing of payload data. The data sharing between BSs costs additional signaling overhead and increases the burden on the backhaul link. In order to alleviate the backhaul capacity requirements in C-RAN, compression strategies and data sharing strategies have been proposed. Specifically, compression strategies performed by the central processors compress and forward the beamformed signals to the BSs after beamforming. Data sharing strategies take advantage of the association between

UEs and BS clusters. Then, the central processor only shares messages among UEs and its serving BS clusters. Moreover, BSs need to transmit the signals to the UEs cooperatively after beamforming.

- UE-centric cluster
  The performance of the data-sharing strategy mainly depends on the selection mode between UEs and BS cooperative clusters. In UE-centric clusters, each UE is served by a separate subset of neighboring BSs. Since a cluster of UEs may overlap with other clusters, UE-centric clusters have no explicit cluster edges. The details of the implementation for UE-centric clustering are discussed in [185].
- Energy saving and green C-RAN
  Compared to the power consumption at BSs, the power consumption of the transmission network can be negligible. Therefore, previous works on the energy efficiency of cellular networks only considered power consumption at BSs. However, the power consumption of the transmission network has a greater impact on the energy efficiency in C-RAN. Moreover, a large number of RRUs do play a positive role in performance improvement for C-RAN, but also lead to excessive interference and energy consumption. As a result, for transmission links and the corresponding RRUs, supporting sleep mode is critical to reducing the power consumption of the C-RAN.

A case study for Re-configurable Backhaul in C-RAN Based Small Cell Networks

In order to achieve the above mentioned data sharing strategies, the backhaul of the C-RAN system which can be reconfigured to adapt to different traffic load conditions is proposed [186]. During the running time of the system, the reconfiguration of the backhaul connection is implemented by the central processing node. The key operations of the central node are described as follows:

- BBU decoupling
  In BBU decoupling, BBU processing is decoupled from route area update (RAU) and UE buffers, and the transmission signals (frames) generated by a given set of BBU units are sent to different sets of RAUs at contrasting times. Moreover, UE data buffers can be decoupled from BBU processing so that multiple BBUs are able to share UE data. Since the decoupling processing can be applied separately during static scenarios and dynamic handover periods, different BBUs can serve the same cell at separate times.
- BBU selection and scheduling
  According to the service requirements of UEs and the assignation requirements of RAU, the BBU selector determines the appropriate number of BBUs to generate the distinct number of transmission frames. The transmission frame is determined by the joint scheduler, by considering the UE buffer and the traffic requirements. Overall, the joint scheduler has strong flexibility to support resource virtualization of the entire community or the entire network.

- Reconfigurable switch
The appropriate transport frames generated after physical-layer processing is transmitted to the correct RAU on the common public radio interface . Since some frames can be sent to multiple RAUs while others are separately sent to the specific RAU, the reconfigurable switch fabric need to have functions of unicast switching and multicast switching. The switch module can activate the input signals based on the expected set of receiver RAUs for the frame. Then, the frames are allowed to replicate across multiple RAUs. Since the BBU pool serves dozens of RRUs, the switch fabric is designed to consist of multiple smaller-sized switches to ensure the scalability of reconfiguration coverage.
- Optical conversion and distribution
The digital radio signal of the switch at the output is converted into the optical signal by loading onto the specific optical wave. The optical signals with different wavelengths are multiplexed by wavelength division multiplexing (WDM). The individual signals with various wavelengths are transmitted to the small cells by separate add/drop multiplexers, respectively. Next, the optical signal is reconverted back to the electrical signal. Finally, the signal modulated to RF achieves transmission on RAUs.

In this work, they provide a case for the re-configurable backhaul, aiming at the realization of UE-centric clustering and energy saving. Specifically, the architecture prototype consists of a WiMAX-based C-RAN with four BBUs and four RAU frames, where the frames are transmitted from BBUs to RAU via backhaul with optic fiber. Through various real-world experimental schemes of WiMAX clients, the advantages of the re-configurable backhaul in meeting UE heterogeneity and traffic requirement can be verified. Moreover, the potential for energy saving in the BBU pool is proved as well.

4.5 MMWAVES

According to the market forecast, wireless data services will grow $500-1000$ times in the next 10 years, with an average annual growth of $1.6-2$ times. Such a development trend poses huge challenges to the service capabilities of 5G [187]. On the one hand, the rising number of smart terminals lead to an explosive growth of wireless network traffic. On the other hand, the abundance of applications promote transmitted data beyond traditional data type, e.g., texts or images. Compared with the texts and images, the transmission demands of mobile TV and other multimedia services (e.g., audio, video) are taking up an increasing proportion of the communication system. Therefore, 5G must expand the network capacity effectively to meet the increasing demand of traffic in quantity and type.

Although multiple technologies can achieve the capacity expansion of a wireless communication system, the adoption of higher frequency bands is the easiest and the most direct method. In fact, the theory that supports the above capacity improvement scheme is derived from Shannon's theorem: $Wlog(1+S/N)$, where W is the

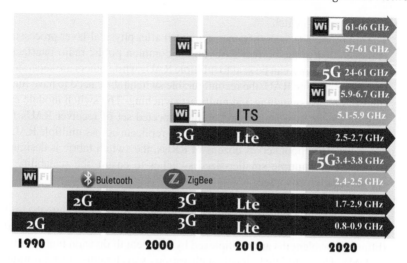

Figure 4.7 Wireless system spectrum evolution

allocated spectrum bandwidth of the channel, S is the average signal power, and N is the average noise power. Since the channel capacity is proportional to the spectrum bandwidth, the application of high-frequency bandwidth can meet the large capacity requirement. At present, the common frequency bands for mobile communication systems are mainly below 3 GHz because of their better wireless propagation characteristics. However, the spectrum resources with such frequency bands are finite, the available spectrum resources are extremely scarce. In order to effectively expand the spectrum capacity, the frequency bands above 3 GHz (e.g., mmWaves) have drawn much attention, since they can provide a lot of available spectrum resources. The preliminary prediction shows that the available spectrum resources in the mmWaves band are 200 times more than those in the band below 3 GHz. Figure 4.7 shows the evolution of the spectrum in the wireless communication system.

mmWaves Propagation Characteristics

The frequency range of mmWaves is from 30 GHz to 300 GHz. Then, the wavelength range of mmWaves is from 1 mm to 10 mm. According to recent research, mmWaves with the frequency between 57 GHz and 64 GHz can be affected by oxygen absorption, while mmWaves with the frequency between 64 GHz and 200 GHz are affected by water vapor absorption. Therefore, the mmWaves with the mentioned spectrum range are suitable for short-distance communications. However, the remaining spectrum resources have similar propagation characteristics. Thus, they can be applied to mobile broadband communication with large coverage [188].

- Penetration loss
  Penetration loss is a critical issue in wireless communications, which affects the coverage of outdoor BSs. Compared with the frequency band below

3 *GHz*, mmWaves experience significant penetration loss when penetrating buildings or obstacles. Since the penetration loss can reach hundreds of **dB**, outdoor BSs using mmWaves can hardly provide services for indoor UEs. Therefore, deploying WiFi or other indoor LPNs is an effective solution to solve the above problems.

- Multipath
  The propagation of mmWaves in the actual wireless communication environment is affected by reflection and diffraction. Multipath formed by reflection and diffraction will further limit the propagation range of mmWaves. Therefore, mmWaves are generally used in non-line-of-sight transmission scenarios. The reason is that the mmWave broadband communication system can generate narrow beams through multiple antennas to decrease multipath components. According to the actual measurement for the system with bandwidth between 10 *MHz* and 100 *MHz*, the multipath delay of mmWave in the urban environment is only $1 - 10$ *ns*.

- Doppler frequency shift
  Doppler frequency shift depends on the moving speed of UEs and the frequency of the working carrier. For example, in the mmWave communication system with rich scattering, if a UE moves with the speed increasing from 3 *km/h* to 350 *km/h*, the range of Doppler frequency shift is 10 *Hz*−20 *KHz*. Note that the Doppler frequency shift of receivers in different directions are different. Then, it can form a larger Doppler frequency spread range. However, mmWave communication system with large coverage always generates narrow beams through multiple antennas. Thus, the Doppler spread range can be effectively reduced.

A Combination of mmWaves, Massive MIMO, and UDNs

The mmWave communication system mainly consists of multiple BSs to achieve mmWave signal transmission and reception. The antennas used in mmWave communications have the advantages of small size and easy mass production. Therefore, these antennas can be effectively used in system densification to achieve the goal of coverage improvement [189]. Moreover, the combination of mmWave technology, UDN, and massive MIMO has shown greater potential. UDN shortens the communication range of mmWaves, while massive MIMO further extends the propagation distance of mmWaves. Although such a combination seems like a promising development direction, some key challenges still need to be solved in the actual deployment.

- Channel estimation
  If CSI can be accurately predicted in the communication system, then the BSs can reduce interference by multi-UE downlink beamforming, thereby improving the system spectrum efficiency. However, accurate CSI feedback from the UE to the BS requires excessive overhead in large antenna array systems, which poses a huge cost challenge to the actual deployment. Considering that downlink beamforming only has limited CSI feedback, a

feasible scheme is proposed to select random (orthogonal) beams in a rich scattering environment [190, 191].

- Hybrid beamforming & antenna structure
  MIMO precoding/combination in mmWave communications is more complicated than that with lower frequencies. The main reason is that the hardware specifications are not uniform. Since the antenna size in mmWave communications is small, mixed-signal components such as high-resolution analog-to-digital converters (ADC) are difficult to deploy on extremely small antennas. Moreover, the application of massive-MIMO increases the number of antennas. The increased number of antennas affects the complexity of critical signal processing functions, e.g., channel estimation, precoding, combining, and equalization. Therefore, the traditional MIMO transceiver architecture is not suitable for mmWave communications. Currently, a feasible method is to apply hybrid analog/digital precoding and combination schemes or low-resolution ADC schemes to reduce the power consumption of the receiver.
- Multi-UE concurrent beamforming
  Compared with traditional wireless communication in $2.4/5$ GHz band, mmWave communications have higher attenuations. Therefore, techniques, such as directional antennas and beamforming, are wildly used to compensate for the free space propagation loss of communication links. On the one hand, the deployment of directional antennas on the transmitter and receiver enables the directional transmission links to have higher gain and less interference, thereby improving the transmission range, spatial multiplexing rate and throughput. On the other hand, link scheduling algorithms are proposed to optimize beamforming technology, thereby improving system throughput [192–194].
- Compressive sensing for beamforming
  Generally, the beamforming of a mmWave system is different from that of a microwave system because mmWave channels suffer from high path attenuation. Furthermore, since there are few paths for reliable transmissions, beamforming based on directional beams needs to utilize the reliable paths in a limited scattering environment. Moreover, hybrid beamforming can achieve superior performance with limited hardware complexity. In order to further optimize the performance of beamforming, compression sensing is introduced to estimate sparse signals and parameters from measurements [114].
- 3D beamforming
  Most of the existing channel models are two-dimensional, these models only consider the wave propagation in the azimuth plane. However, the actual waves propagation is affected by the channel components in the elevation direction. Accordingly, three-dimensional beamforming for full-scale MIMO is proposed. To make full use of the additional freedom degrees of the three-dimensional channel, the model must adapt to the beam pattern of

each UE in the vertical direction. Thus, the signal strength at the receiver can be increased while the interference to other UEs can be alleviated.

UDNs in Millimeter-wave Frequencies

A vital component of 5G wireless access is UDN which provides the extremely high data rate for localized and low-mobility cells [195]. Specifically, UDN can increase the data rate to 10 GB/s and reduce the delay to 1 ms in the localization environment. Moreover, UDN allows large continuous bandwidth up to about 2 GHz in the mmWaves band. In order to further reduce the need for physic connections between access nodes and support networks, access nodes in UDNs can utilize wireless self-backhauling to achieve the connection with wired aggregation nodes.

Wireless self-backhauling can improve accessibility and coverage by simplifying the connections among access nodes. Specifically, access nodes in UDNs connect to aggregation nodes, and aggregation nodes connect to high-bandwidth transmission networks. Communication data must go through the operator's data center before it reaches the core network. Therefore, the scale of the network should be consistent with the number of aggregated nodes in the system. Since each hop of wireless backhaul can increase end-to-end delay, the hop count is set as small as possible in typical networks. Commonly, the data from UEs can be transmitted to an aggregation node by two or three hops.

The routing algorithms in wireless networks are different from those in wired networks. Routing metrics associated with a certain link do not need to be independent with other links. Therefore, the suboptimal algorithms with reasonable complexity have better performance. Generally, the optimal verification for a single metric is simple. However, considering the joint criteria, such as low latency and high throughput, the verification of metrics becomes difficult. A suboptimal solution is to establish routes that only optimize throughput while ignoring latency, then prune links that are a bit smaller than the maximum throughput. Then, the route with the best delay can be produced in the second stage.

Mobile wireless access systems must operate in the defined dedicated spectrum, which is mainly composed of licensed spectrum. However, with the introduction of spectrum sharing technology, the dedicated system bandwidth of about 2 GHz has become difficult to carry the access load of multiple parallel networks. Spectrum sharing in the frequency range of mmWave is feasible for UDNs due to the advantages of mmWave on the low power and the wide deployment range. Moreover, spectrum access based on the principle of license shared access (LSA) is another effective solution. Since the access range of LSA is small, UDN scenario is suitable for this technology to reduce the overall interferences. The premise of adopting LSA is to acquire the interface of the external spectrum repository. Then, UDNs can utilize this interface to obtain information about available frequencies and related operational constraints.

4.6   PROACTIVE CACHING

Network densification reduces the service range of the cells and further improves the QoS of the communication system. Meanwhile, the dense deployment of cells results in the growth of backhaul load, thereby deteriorating network performance inevitably. Caching is widely regarded as an effective way to significantly reduce the burden on backhaul links. The basic idea of caching is to dynamically store UE-required files. Since UEs can obtain the required files directly from the BSs without connecting to the core network, caching schemes can effectively reduce transmission delay and network congestion [196].

Currently, mobile network caching models are built based on reactive caching. When a UE service request is proposed, the content servers will immediately provide the corresponding service. However, content servers in reactive caching schemes are difficult to respond timely, and network performance will decrease with the surge in service requests. Although the ideal solutions for traditional network models are expanding the buffer space of BSs while improving the performance of mobile backhaul links, financial and technical constraints make it impossible to alleviate the huge pressure on network traffic fundamentally. Active caching is an excellent solution to reduce the peak traffic of mobile backhaul links while improving QoE. In contrast to reactive caching, active caching can more intelligently manage mobile networks by predicting UE behavior preferences with historical behavior.

The Location and The Contents of The Proactive Caching

Proactive caching enables UEs to obtain the requested content from small cells or other devices. Since UEs do not need to obtain content from content service providers through the core networks and wired networks, the imbalance between wireless demand and available capacity can be improved. Typically, proactive caching includes two steps, which are content placement and content delivery. The content of the cache, the location of the cache, and the way to download the content should be determined in the content placement step. In the content delivery step, the content delivering to the requesting UE should be determined. The above two steps are executed under different conditions. When network traffic is low and network resources are cheap and abundant, content placement will be performed. On the contrary, when network traffic is high and network resources are scarce and expensive, content delivery will be performed. Currently, the focus of the existing research in active caching mainly concentrates on the location and content of the cache.

- Cache location
  Macro BS: Since the macro BSs have a larger coverage area than the small BSs, they can provide services for more UEs. Moreover, when the backhaul link resources are scarce, the cache on the macro BSs can achieve higher cache hit rates and lower deployment costs.
  Small BS: Small BSs in heterogeneous UDNs are closer to the UE terminals than the wireless BSs in conventional cellular networks. Therefore,

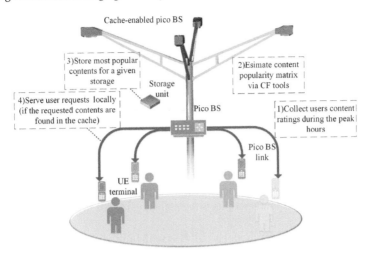

Figure 4.8 A practical procedure for proactive caching at the BSs [197]

deploying cache at small BSs can provide faster responses. A sketch of the proactive caching procedures at the BSs is shown in Fig. 4.8.

Mobile device: The development of D2D communication technology has promoted the generation of UE-side cache solutions. On-demand fetching can be achieved by distributing cached content on mobile devices, thereby greatly meeting the QoS requirements of UEs.

• Cache content

Popularity of files: Since popular content has a higher frequency of access, researchers begin to focus more on the popularity of cached files. Actually, the popularity of a cache file refers to the UE requestion probability that a file in the file library of a certain area is requested by most UEs. According to literature [198], the popularity of the content is subject to the Zipf distribution, which can be represented by the parameters about the size and popularity of a file library. Generally, the changes in content popularity distribution are slower than the traffic changes in the cellular network. Meanwhile, the content popularity distribution usually remains roughly the same for a lengthy period. A similar example is that the popularity of movies is usually 1 week, while the popularity of messages is usually 2 or 3 hours [199]. Moreover, it is worth noting that the popular content in a large area such as a city or a country is often different from that in a small area such as a campus.

UE's preference for content: Another factor related to content caching is UE preference, which makes a difference in the probability that a specific UE requests a file within a certain time. Since UE often has a strong preference for content in a particular category, caching such content can increase the cache hit rate. In order to improve the average cache hit rate, recommendation algorithms can be used in content preferences prediction. Specifically, the content without matching the target can be filtered out based on the historical data requested by UEs [200].

Proactive Small Cell Networks

Small cells are usually deployed in the areas where the backbone networks are difficult to be constructed. To improve the reliability of backhaul transmissions, signals in backhaul links are usually transmitted by copper and mmWaves. However, it is still a challenge for the capacity restriction the backhaul links. Active caching is regarded as a potential solution to alleviate the backhaul burden. For example, Bastug *et al.* [201] propose a proactive caching scheme to store files according to the highest popularity of files. This method reflects the mainstream idea in current active caching schemes.

In the proposed network model, small BSs with limited capacity in backhaul links are deployed with mass storage units. Meanwhile, small BSs can grasp the information of the popularity matrix $\mathbf{P}_{N \times F}$, where each row represents a different UE and each column reflects a different file preference. Considering the actual situation that the popularity matrix is large, sparse, and partially unknown, a distributed proactive caching process is proposed based on supervised machine learning and collaborative filtering. This scheme is inspired by the Netflix paradigm, in which UE file correlation is used to infer the probability that the UE u requests the file i.

The caching process consists of training and placement. The goal of training is to accurately estimate the popularity matrix $\mathbf{P}_{N \times F}$ for each small BS. The models of UE preference based on existing information can be optimized by the minimization of the following least-squares problem,

$$\min_{b_u, b_i} \sum_{u,i} (r_{ui} - \hat{r}_{ui})^2 + \lambda \left(\sum_u b_u^2 + \sum_i b_i^2 \right) \tag{4.6}$$

Note that the sum is over the user-file pair (u, i) in the training set, and the minimization is over the $N + F$ parameters. Specifically, UE u determines the preference of file i. Term N represents the number of UEs while term F represents the number of files in the training set. Moreover, $\hat{r}_{ui} = \bar{r} + b_u + b_i$ is the baseline predictor, where b_i and b_u are related to the average value $\bar{r}$, reflecting the quality of each file b_i and the quality of each UE, respectively. In order to achieve the balance between regularization and fitting training data, the weight λ should be adjusted carefully. After obtaining the file popularity matrix $\mathbf{P}_{N \times F}$, each small BS will store the files with high popularity greedily in the placement step.

4.7 SUMMARY

In this section, we introduce enabling factors and emerging technologies of UDNs, including massive-MIMO, machine learning, SDN, C-RAN, mmWave, and proactive caching.

MIMO is an effective technique in improving space diversity, which can increase system capacity and reduce interference. A reasonable combination of UDNs and MIMO can improve system capacity and transmission rate. However, with the increased utilization of high-band spectrum resources, signal attenuation becomes more serious. Nowadays, MIMO has gradually expanded into massive-MIMO, which can improve signal coverage by generating beams with a higher gain. In general, massive-MIMO further eliminates interference among UEs, and reduces latency and deployment costs. Thus, spatial resolution and array gain can be greatly improved. However, massive-MIMO also faces challenges in design and implementation, such as pilot pollution, difficulty in obtaining global CSI, and expensive hardware costs. At present, a feasible idea is to combine massive-MIMO with mmWave and UDNs to improve the reliability of the backhaul. Specifically, massive-MIMO enables mmWave to propagate farther, while UDN shortens the propagation distance of signals, meanwhile, mmWave makes antennas smaller and more suitable for large-scale manufacturing.

Machine learning, as a statistical knowledge-based technique, enables the system to learn from data autonomously, hence, it is often used to solve optimization problems. The utilization of machine learning makes 5G wireless networks have high-precision predictability and strong initiative. Therefore, the network can be sliced based on different application scenarios and business requirements. Machine learning is further divided into supervised learning, unsupervised learning, and reinforcement learning. Specifically, supervised learning, including regression models, KNN, SVM, and Bayesian models, can effectively solve the problems of channel noise estimation, best handover selection, and spectral feature estimation in UDNs. Unsupervised learning, including K-means clustering, PCA, and ICA, can effectively solve the problems of clustering, anomaly detection, fault detection, and intrusion detection problems. Furthermore, reinforcement learning, including MDP, Q-learning, and MAB, can provide decision support for system resource allocation, interference management, and other issues in UDNs.

SDN is an emerging network architecture with an independent structure of control and forwarding, which has centralized control logic and opened network functions. It breaks the hardware limitation in the network architecture, thereby achieving the goal of flexible network modification and rapid application deployment. The densification of UDNs cannot be completely achieved by readjusting current protocols. Hence, SDN is undoubtedly the best choice. CROWD is a management structure optimized for energy-intensive wireless dense networks. This scheme uses a two-layer dynamic SDN controller structure to optimize the aggregation of control information and reduce signaling overhead. Another management architecture is WiSEED, which manages a set of operational services to make network systems more scalable, flexible, and energy-efficient by SDN.

C-RAN is a green wireless access network architecture based on centralized processing, cooperative radio, and real-time cloud infrastructure. C-RAN forms a centralized dense network through the dense deployment of RRUs. Moreover, C-RAN has great advantages in terms of cost, capacity, and flexibility in networks. C-RAN also provides UEs with better services than UDNs in some scenarios. The current research on C-RAN mainly focuses on decreasing the backhaul capacity requirements, accurately implementing UE-centric clustering, and reducing system transmission power consumption. In a C-RAN based reconfigurable backhaul architecture, the central node needs to decouple the BBU processing from the RRU and UE buffers. Then, BBU selection and scheduling are implemented. After that, reconfigurable switching is implemented. Finally, optical conversion and signal distribution can be achieved. The proposed architecture is built on a WiMAX-based C-RAN system. Specifically, this architecture not only meets the traffic requirements of heterogeneous UEs, but also saves significant energy for the system.

The scarcity of spectrum resources causes increased attention on mmWave in the communications field. The frequency range of mmWave is between 30 GHz and 300 GHz. Therefore, mmWave usually has a large attenuation loss and is prone to reflection and diffraction. These characteristics determine that mmWave is only suitable for short-range communication. In order to effectively implement mmWave communications, massive-MIMO can be used to expand the transmission range. However, the combination of massive-MIMO and mmWave still faces challenges in channel estimation, multi-UE parallel beamforming, 3D beamforming, compressive sensing, etc. Using the sufficient spectrum resources of mmWave, UDNs can provide better network services to UEs. We give a mmWave application example in the office environment. OFDM is used to implement the physical layer design, wireless self-backhauling is used to achieve the connection of access nodes to wired aggregation nodes, and beamforming is used to reduce interference, thereby greatly improving the backhaul throughput of the network.

Caching is an effective technique to dynamically store files that are often required by UEs in BSs or UE equipment, to significantly reducing backhaul link loss. Different from the current responsive caching model in mobile networks, proactive caching can accurately predict from the UE's history behaviors, so that UE can obtain the requested content from small cells or other devices, improving QoE while reducing the peak traffic of the mobile backhaul link. The content placement in proactive caching is concentrated on macro BSs, small BSs, and mobile devices. The caching content is selected based on the popularity of the files or the UE's preferences. In general, the current proactive caching strategies are divided into the following three categories: machine learning-based, collaboration-based, and UE mobility-based. In the case of cell network transmissions with proactive caching based on the popularity of files, we conduct a detailed analysis of the content training and content placement.

5 Promising Applications

5.1 INTERNET OF THINGS

Internet of things (IoT) interconnects numerous machines, analytics sectors, and people at work, to revolutionize industrial process [202]. IoT realizes real-time data processing and information access while building device connection links. In order to further expand the advantages of IoT, technologies such as smart sensors, wireless communication [203], context-aware computing, cloud computing, and Internet protocols are introduced as building blocks in IoT architecture. In particular, the machine-to-machine communication method and relative standardization works can play vital roles in the development of IoT [204].

IoT not only enables physical objects to see, hear, and think, but also allows them to share information and coordinate decisions. In fact, these objects are transformed into intelligent components by underlying technologies of IoT, thereby having the ability of full precision and accurate decision-making. Specifically, the application domain-specific services (i.e., vertical markets) are constituted by smart objects and their assumed tasks, while the application domain-independent services (i.e., horizontal markets) are constituted by ubiquitous computing and analysis. Figure 5.1 illustrates the overall concepts of IoT, where each domain-specific application interacts with services in separate domains, while sensors and actuators can directly communicate in each domain.

UDN: Enabler for The IoT

The actual deployment of IoT faces great challenges. First of all, trillions of things or devices will connect to the network. The increment of network traffic brings pressure on network management. How to support the large scale of device connection becomes a key issue. Moreover, IoT applications such as healthcare systems, transportation systems, smart grids, and public safety emergency response systems require real-time information. Once urgent events such as health emergency, traffic accident or congestion occur, the relative message needs to be transferred immediately. Therefore, how to improve the performance of wireless networks to support diverse types of applications becomes another principal issue.

UDN technique solves the above problems by providing ambient communication services. First, BSs being densely deployed in UDNs can increase network capacity, thereby improving the scalability of IoT applications. Moreover, in UDNs, each BS has sufficient processing ability. Then, the requirements of low latency and real-time services can be satisfied effectively. Meanwhile, communications under various protocols are compatible in UDNs. It means different applications can share the same network. According to the above description, UDN has the advantages of

DOI: 10.1201/9781003148654-5

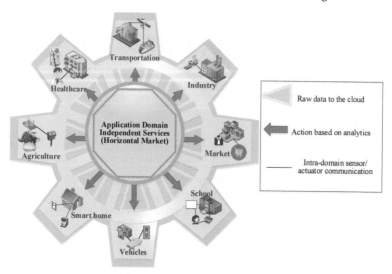

Figure 5.1   The overall picture of the IoT emphasizing the vertical markets and the horizontal integration between them

high network capacity, high real-time performance, and high compatibility. However, there are still challenges in using UDN technique to support IoT applications, which are given as follows:

- Deployment
  With billions of embedded devices swarming into the cellular networks, the deployment of UDN becomes complex. How to deploy small BSs macro BS is acknowledged as a main foreseeable challenge of IoT [205].
- Asymmetry uplink/downlink traffic demand
  Since IoT devices are mainly focusing on data uploading, the traffic ratio of uplink and downlink is different from that of traditional communications [206]. The uplink traffic is usually much more than downlink traffic in IoT.
- Energy consumption
  The new network paradigm needs to meet the requirements of IoT function. However, some IoT devices, such as smart meters, need to operate in low-power mode since their limited battery capacity [207, 208].
- Hybrid network demands
  In addition to the mobile broadband services, future cellular networks should cope with a range of new IoT services such as uRLLC, mission critical applications, etc [209].
- Reliability and delay
  IoT requires real-time feedback information in applications such as remote e-healthcare, smart homes, environmental monitoring, and industrial automation. Then, high reliability and low delay are critical conditions to guarantee such application scenario [210].

5.2   MOBILE EDGE COMPUTING

MEC for UDN: From Clouds to Edges

MEC is considered as a promising technology for the improvement of UDN services. Specifically, MEC allows computing tasks to be performed at the edge of the network, thereby ensuring the effective execution of compute-intensive and latency-critical applications on resource-constrained mobile devices. The key components of the MEC system are shown in Fig. 5.2. MEC servers play the roles of small data centers in UDNs. These servers are really close to the UEs and can coexist with access points (e.g., BSs). Generally, the server connects to the data center via the gateway and Internet backbone. In order to establish reliable wireless links in the MEC system, advanced air interfaces are introduced to separate the server from the mobile device.

- Collaborative offloading
  Figure 5.3 shows collaborative offloading strategies including MEC server offloading, D2D offloading, cloud-edge collaborative offloading, and subtasks offloading. In literature [211], the authors proposed a computing task offloading strategy to minimize the overall energy consumption of the mobile terminal under the constraints of execution delay.
- Optimization
  An optimization can be formulated to minimize the weighted sum of execution delay and device energy consumption during MEC offloading. Joint optimization between load scheduling and transmission power allocation in MEC systems with multiple independent tasks can be achieved [212]. The authors in [213] propose a task assignment optimization strategy based on graph matching to solve the problems of high-performance D2D group task assignment with constraints. The authors in [214] propose a blockchain-based credit system to achieve task scheduling fairness among UEs in D2D networks, they design a super node for BS to schedule computing tasks collaboratively according to UE mobility and credit balance.

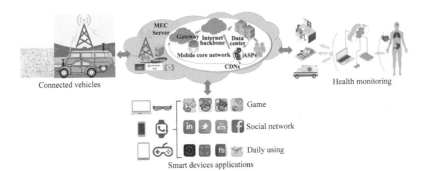

Figure 5.2   Architecture of the MEC systems

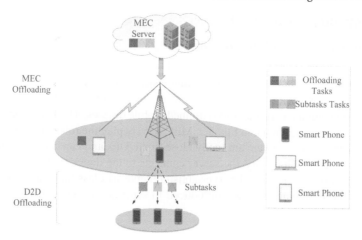

Figure 5.3   Collaborative offloading

- Caching

  The orchestration of computing, communication, and caching also plays a key role in MEC for UDNs. Literature [215] proposes a green wireless network architecture based on the combination of SDN. Since the proposed architecture can realize dynamic resource scheduling between different networks with the support of computing services, the diverse needs of different network services can be satisfied comprehensively. In literature [216], a small cellular network system architecture combined with interference alignment technology is proposed. In order to obtain the optimal interference alignment and server selection strategies to minimize network traffic and energy consumption, an exhaustive search algorithm is introduced to optimize node caching, computation, and bandwidth resources jointly. Therefore, the proposed scheme has advantages of simple topology, high throughput and low backhaul load.

5.3   WIRELESS ENERGY HARVESTING

Since the energy consumption of small cells in idle mode is not negligible, the energy efficiency of the entire network decreases with the network densification. Therefore, increasing the energy efficiency in UDN is one of the key directions for mobile network operators to reduce operating expenses and carbon footprint.

When bestowing wireless devices with energy harvesting capabilities, wireless end nodes can continuously obtain energy from natural or man-made phenomena. In this case, the wireless system can work by self-sustaining for a long time. Wireless energy harvesting has benefits in terms of reducing the usage of conventional energy and reducing carbon footprint. Considering that the corresponding wireless system no longer needs a traditional battery for charging, communication scenarios can cover hard-to-reach places, such as remote rural areas and the human body [191]. Due to the outstanding advantages in unfettered mobility, the wireless energy harvesting technique has been introduced to new applications including medical communication, environmental monitoring, and secure transmission.

Wireless Backhauling for Energy Harvesting in UDNs

Based on energy harvesting, UDNs can eliminate the dependence on the power grid when executing data transmissions in backhaul and access links. Energy harvesting is executed on APs. Thus, this kind of APs is also called autonomous-APs (A-APs). Specifically, A-APs collect energy from solar panels or wind turbines for the operation of backhaul and radio access. Moreover, excess energy is also stored for later use.

The demands of cellular network capacity change dynamically over different periods of a day [217]. Therefore, the mobile network can be flexibly configured to meet the maximum capacity requirements during the peak period. In order to save energy, free APs (i.e., APs that are not responsible for cellular coverage) can be turned off during the low traffic period. Meanwhile, the main APs with fiber backhaul and grid power, named as M-APs, can be configured to provide the backhaul connection for one or more A-APs. Generally, an A-AP needs to cover at least one M-AP. Besides, M-APs can provide services to UEs and A-APs at the same time.

Figure 5.4 shows the basic schematic diagram of A-AP, which is composed of the energy sub-system, backhaul transceiver and access transceiver. Particularly, energy sub-system has an energy harvester, a battery and a power management system. The process of energy transmission from the battery to the backhaul transceiver or access transceiver is controlled by the management system. Apart from energy sub-system, M-AP has a similar set of components like A-AP [218]. Figure 5.5 shows the wireless links among UEs, A-APs and M-APs. A UE can connect to the network through an A-AP access link, an M-AP access link, or a combination of them.

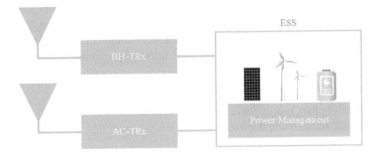

Figure 5.4   A-AP with separated transceivers for backhaul and radio access

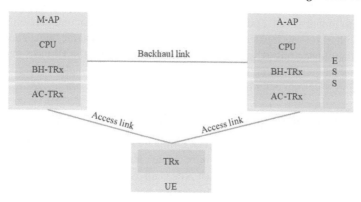

Figure 5.5   The wireless links between M-AP, A-AP, and UE

5.4   SELF-DRIVING OR AUTONOMOUS VEHICLES

Big Data Driven Autonomous Vehicles

With the development of automobile technology, people have more expectations for driving experiences with future cars. Besides factors such as security, environmental protection, and energy saving, the improvement of autonomous driving has also drawn much attention from researchers. Autonomous vehicles can achieve reliable information exchange by vehicle-to-vehicle (V2V) communications and vehicle-to-infrastructure (V2I) communications. Based on the efficient communication methods, recent autonomous vehicle studies begin to focus on the applications of road safety enhancement, traffic management, vehicle mobility data services, and autonomous driving assistance [192, 219].

With the increasing demand for mobile services, the amount of data required, generated, collected, and transmitted by autonomous vehicles increases exponentially. Literature [193] believes that data in vehicular ad hoc networks (VANETs) can well match the "5Vs" of big data features, namely, volume, diversity, speed, value, and accuracy. Actually, future VANETs will rely on big data to enable a variety of promising applications and services such as smart cities and intelligent transportation systems. Meanwhile, the adoption of big data will significantly change all aspects of society, including transportation systems, telecommunications, business, government, and human lifestyle. VANETs enabled applications are shown in Fig. 5.6.

UDN: Enabler for Autonomous Vehicles

The trend of big data brings challenges and opportunities to autonomous vehicles simultaneously. In order to effectively support advanced driving, autonomous vehicles should provide extremely high data rates, large network capacity, heterogeneous network integration, and differentiated QoS guarantee. An intuitive solution to meet the above requirements is pervasive UDNs. According to key performance indicators, UDNs can provide data rates of 10 Gb/s with an end-to-end delay of less than 1 ms.

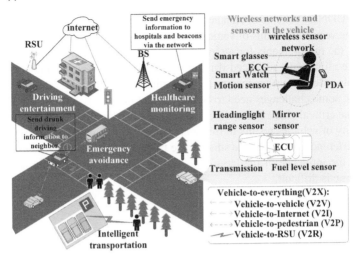

Figure 5.6 VANETs applications

Meanwhile, low power consumption and high-reliability requirements of emerging IoT applications can be satisfied in UDNs as well.

To support different services, three types of use cases with specific performance metrics are defined, including eMBB, URLLC, and mMTC. These three use cases combined with well-defined key technologies guarantee the performance of VANETs big data collection and transmission tasks.

- eMBB
  In autonomous vehicles, large-capacity networks should provide extremely high data rates to cope with the exponential increase demand for automotive mobile big data services. Since eMBB can provide 10 Gb/s of peak data rates and 10 $Tb/v/km^2$ of mobile data volumes, the emerging data-eliminated vehicle data applications in UDNs can be better supported.
- URLLC
  Mission-critical data services in autonomous vehicles require extremely low latency and extremely high reliability. The above requirements in UDNs fall under the purview of URLLC, with a latency of less than 5 ms and reliability of over 99.999%.
- mMTC
  mMTC relies on machine-to-machine communication and potential technologies to support ubiquitous machine-like connections with low energy consumption and low latency. Most data are generated by lightweight devices, such as sensors deployed on vehicles or roads. In order to facilitate big data collection services, devices can be densely deployed in UDNs to accommodate a large number of concurrent connections, thereby improving the data transmission reliability as well.

In UDNs, a scheme of enhanced vehicle-to-everything(eV2X) is proposed to support the vertical domains of vehicle communications and data services [194]. Specifically, the requirements for a typical V2X scenario are defined in UDNs, including vehicle alignment, advanced driving, extended sensors, and remote driving. However, in order to further inspire the potential of autonomous vehicles supported by UDNs, some challenges need to be addressed first.

One critical challenge is time-varying network topology and channel state. High mobility of vehicles results in frequent changes in the network topology. In addition to interference caused by the small-scale fading, the channel state is greatly changed under the influence of the large-scale fading. Therefore, capturing accurate CSI in real-time is difficult.

5.5 SUMMARY

In this section, we discuss some promising applications in 5G communications, including IoT, MEC, wireless energy harvesting, and autonomous driving.

IoT devices connected to the network can share information and communicate directly without human intervention, thus, intelligent decision-making can be achieved. Moreover, smart sensors, wireless communication technologies, context-aware computing, cloud technologies, and Internet protocols are the foundations of the IoT architecture. There are still many challenges when using IoT applications in UDNs. For example, large-scale and scalability, asymmetric traffic requirements, excessive energy consumption, mixed traffic type requirements, and ultra-reliability/ultra-low latency requirements all need to be solved in the development of IoT.

The application of MEC techniques enables computational-intensive and latency-critical applications to be performed on resource-constrained mobile devices. Besides, the application of MEC can also effectively reduce costs in mobile energy consumption, network traffic, and transmission link. The key components of a MEC system include mobile devices and MEC servers, in which MEC servers are generally close to the terminal UEs and can coexist with wireless APs. Research on MEC mainly focuses on the seamless integration of wireless communication and mobile computing, while ultra-dense edge computing needs to orchestrate a variety of resource management strategies, such as offloading, optimization, and caching.

The energy harvesting technology enables wireless devices to have energy harvesting functionality and improves the self-sustainment level of wireless networks. Thus, it can reduce conventional energy consumption and the corresponding carbon footprint. In an energy harvesting network, the AP can perform backhaul and access link data communication at the same time. Hence, in UDNs, the dependence of the optical distribution network and the power grid can be eliminated. Moreover, the harvested energy can be stored for backup. A-APs can use the collected energy to perform operations of backhaul and wireless access. Considering that the capacity requirements of a cellular network change over time, BS sleeping can be employed during low traffic periods to save energy.

Autonomous vehicles can achieve efficient and reliable information exchange through V2V and V2I communication. However, with the growing demand for mobile services and the rapid development of self-driving technology, the amount of data required for autonomous driving systems has shown an exponential growth trend. The advanced driving has brought great challenges to autonomous vehicles, such as high data rates, large network capacity, heterogeneous network integration, and differentiated QoS guarantees. To address these challenges, UDNs are applicable solutions. UDNs presents three use cases of eMBB, URLLC, and mMTC to ensure the performance of the VANETs. In addition, UDNs also show eV2X use cases for vertical domains that support vehicle communication and data services. However, some challenges also need to be addressed to implement UDN-supported self-driving, such as highly time-varying network topology and channel status.

6 Summary and Future Work

The explosive growth of traffic poses huge challenges to the current wireless communication networks. Constantly emerging technologies accelerate the evolution of ultra-dense networks (UDNs) in 5G networks and even 6G networks. UDNs, as inevitable development trends of future networks, provide UEs with high QoE while meeting the capacity and performance requirements of the system. In this book, we firstly introduce the network evolution and the development trends of future networks, then explain the concepts and characteristics of UDNs. In addition, we conduct a detailed discussion of modeling technologies, resource management, interference management, and mobility management issues involved in UDNs. Then, we discuss some enabling factors of UDNs and emerging technologies related to UDNs, which can improve the performance of the network system. At last, we introduce several promising network applications for UDNs with expanded network functions and greater advantages. Through the above contents, we not only have a general comprehension of UDNs, but also understand the fundamental differences between UDNs and traditional heterogeneous cellular networks. As an important driving factor for the future development of the network, the technologies and extended applications involved in UDNs still need to be further studied. In the final part of this book, we will discuss some challenges associated with existing UDN solutions and outline future research directions.

The resource management of UDNs becomes highly complex since multi-dimensional resources need a joint assignment in numerous irregular and overlapping small cells. Although we mentioned some methods based on stochastic geometry and stochastic optimization in the previous sections, the resource optimization in UDNs still faces high complexity. The second challenge is the high signaling overhead in UDNs. In order to effectively implement the optimal resource allocation strategy, the system needs to obtain the timely status information of the entire network. Note that a large amount of information interaction will bring a huge overhead. Applying distributed structure in the small-scale network can effectively reduce latency and overhead, but it is not applicable in large-scale network scenarios. In light of this, we need to explore large-scale network decomposition and AI technology, thus achieving an accurate prediction while reducing the system overhead. Finally, UDNs also face serious energy consumption. Although the energy consumption of a single small BS is low, the total power consumption of the whole network with massive devices is high. It is known that the evolving networks are random in nature, so the random behavior can be studied in depth through stochastic geometry and random matrix theory. Thus, the reasonable deployment schemes of the BSs can be designed. Moreover, with the continuous development of communication networks, traditional cell sleeping strategies are no longer sufficient for high-density BS systems. Therefore, on the one hand, we need to consider BS cooperation strategies for more effective energy-saving. On the other hand, we need to consider different sleeping and wake-up modes, so that the system can assign different sleeping modes to different cells based on power consumption levels, achieving the purpose of energy saving better.

For interference management, the primary challenge of UDNs is the high computational complexity. The dense deployment of UDNs shortens the distance between UEs and BSs, making the system more susceptible to interference. On the other hand, interference management is often combined with resource management, including frequency, time, space, caching, and computing resource domains, which will face the challenge of dimensional curse in a collaborative mode. In order to solve the above problems, actively shutting down the lightly loaded BSs with large interference can reduce interference. In addition, a multi-domain interference control scheme can help the system effectively eliminate interference in different resource domains. Apart from the complexity problems, the design of spectrum sharing scheme is also a major challenge in interference management. In addition to deepening the research on existing spectrum sharing strategies for multiple operators, we should also consider multiple resource dimensions to dynamically and flexibly maximize network capacity and minimize interference.

Due to the small and irregular coverage area of cells in UDNs, UEs will generate super-frequent handovers during user mobility. To ensure the seamless connection between heterogeneous UEs and applications in UDNs, effective collaboration-based solutions are needed to address these unique problems. Moreover, propagation modeling is also a major challenge for mobility management. None line of sight transmission is common in office environments, however, the distance between transmitter and receiver is shortened in the dense network, which causes none line of sight transmission to be converted to line of sight transmission with a higher probability. Therefore, a path loss model combining line of sight and line of sight transmissions is more suitable for the increasingly complex UDN environment, and it is also of future research.

Besides the aforementioned issues, UDNs also face challenges in multi-technology convergence. The densification of the network can increase the network capacity by 10–100 times, the key of which is to combine UDNs with advanced technologies such as mmWave and massive-MIMO. With the massive MIMO technique, BSs are equipped with a massive antenna array to achieve high space gain. With the mmWave technique, the communications system can obtain a large number of spectrum resources. Although mmWave is greatly affected by path loss and, it is suitable for short-distance communication. The dense deployment of cells in UDNs creates conditions for the application of mmWave. Multiple advanced technologies can play roles in reducing interference and improving performance through coordination and collaboration. Hence, the perfect and practical integration of these technologies still needs further research from industry and academia.

References

1. Y.-N. R. H. Peng, Y. Xiao, and Y. Yifei, "Ultra dense network: Challenges enabling technologies and new trends," *China Commun*, vol. 13, no. 2, pp. 30–40, 2016.

2. "Cisco visual networking index: Global mobile data traffic forecast," 2018.

3. M. Katz, P. Pirinen, and H. Posti, "Towards 6g: Getting ready for the next decade," in *2019 16th International Symposium on Wireless Communication Systems (ISWCS)*, 2019, pp. 714–718.

4. T. Huang, W. Yang, J. Wu, J. Ma, X. Zhang, and D. Zhang, "A survey on green 6g network: Architecture and technologies," *IEEE Access*, vol. 7, pp. 175 758–175 768, 2019.

5. T. S. Rappaport, Y. Xing, O. Kanhere, S. Ju, A. Madanayake, S. Mandal, A. Alkhateeb, and G. C. Trichopoulos, "Wireless communications and applications above 100 ghz: Opportunities and challenges for 6g and beyond," *IEEE Access*, vol. 7, pp. 78 729–78 757, 2019.

6. X. Bai, Q. Li, and S. Tao, "Resource allocation based on dynamic user priority for indoor visible light communication ultra-dense networks," in *2018 IEEE 18th International Conference on Communication Technology (ICCT)*, 2018, pp. 331–337.

7. S. J. Nawaz, S. K. Sharma, S. Wyne, M. N. Patwary, and M. Asaduzzaman, "Quantum machine learning for 6g communication networks: State-of-the-art and vision for the future," *IEEE Access*, vol. 7, pp. 46 317–46 350, 2019.

8. J. Zhu, M. Zhao, and S. Zhou, "An optimization design of ultra dense networks balancing mobility and densification," *IEEE Access*, vol. 6, pp. 32 339–32 348, 2018.

9. B. Zong, C. Fan, X. Wang, X. Duan, B. Wang, and J. Wang, "6g technologies: Key drivers, core requirements, system architectures, and enabling technologies," *IEEE Vehicular Technology Magazine*, vol. 14, no. 3, pp. 18–27, 2019.

10. D. Kalbande, Z. khan, S. Haji, and R. Haji, "6g-next gen mobile wireless communication approach," in *2019 3rd International conference on Electronics, Communication and Aerospace Technology (ICECA)*, 2019, pp. 1–6.

11. H. Wei, H. Luo, M. S. Obaidat, and T.Wu, "An active updating strategy for caching periodic data in the internet of things," in *2018 IEEE International Conference on Communications (ICC)*, 2018, pp. 1–6.

12. M. Moltafet, N. Mokari, M. R. Javan, H. Saeedi, and H. Pishro-Nik, "A new multiple access technique for 5g: Power domain sparse code multiple access (PSMA)," *IEEE Access*, vol. 6, pp. 747–759, 2018.

13. L. Ma, X. Wen, L. Wang, Z. Lu, and R. Knopp, "An sdn/nfv based framework for management and deployment of service based 5g core network," *China Communications*, vol. 15, no. 10, pp. 86–98, 2018.

14. G. Gu and G. Peng, "The survey of gsm wireless communication system," in *2010 International Conference on Computer and Information Application*, 2010, pp. 121–124.

15. E. Wang, S. Zhang, and Z. Zhang, "Research and implement of ppp and tcp/ip protocol based on gprs," in *2011 7th International Conference on Wireless Communications, Networking and Mobile Computing*, 2011, pp. 1–4.

16. M. kaur, R. K. Chechi, G. C. lall, and Bhawna, "Quality of service approach in homogenous network (umts-umts) using parameter throughput, jitter and delay," in *2012 7th International Conference on Computing and Convergence Technology (ICCCT)*, 2012, pp. 11–14.

17. J. Lu, L. Xiao, Z. Tian, M. Zhao, and W. Wang, "5g enhanced service-based core design," in *2019 28th Wireless and Optical Communications Conference (WOCC)*, 2019, pp. 1–5.

18. M. Jaber, M. A. Imran, R. Tafazolli, and A. Tukmanov, "5g backhaul challenges and emerging research directions: A survey," *IEEE Access*, vol. 4, pp. 1743–1766, 2016.

19. S. Lins, P. Figueiredo, and A. Klautau, "Requirements and evaluation of copper-based mobile backhaul for small cells lte networks," in *2013 SBMO/IEEE MTT-S International Microwave Optoelectronics Conference (IMOC)*, 2013, pp. 1–5.

20. K. O. Lee, J. H. Hahm, and Y. S. Kim, "Pstn/isdn/xdsl evolution to broadband convergence network," in *The 9th International Conference on Advanced Communication Technology*, vol. 3, 2007, pp. 1987–1990.

21. C. Lim, C. Ranaweera, E. Wong, and A. Nirmalathas, "Design and planning for fiber-based small cell backhauling," in *2018 20th International Conference on Transparent Optical Networks (ICTON)*, 2018, pp. 1–3.

22. A. Ting, D. Chieng, K. H. Kwong, I. Andonovic, and K. D. Wong, "Dynamic backhaul sensitive network selection scheme in lte-wifi wireless hetnet," in *2013 IEEE 24th Annual International Symposium on Personal, Indoor, and Mobile Radio Communications (PIMRC)*, 2013, pp. 3061–3065.

23. L. Dussopt, O. El Bouayadi, J. A. Z. Luna, C. Dehos, and Y. Lamy, "Millimeter-wave antennas for radio access and backhaul in 5g heterogeneous mobile networks," in *2015 9th European Conference on Antennas and Propagation (EuCAP)*, 2015, pp. 1–4.

24. H. Zhang, Y. Dong, J. Cheng, M. J. Hossain, and V. C. M. Leung, "Fronthauling for 5g lte-u ultra dense cloud small cell networks," *IEEE Wireless Communications*, vol. 23, no. 6, pp. 48–53, 2016.

25. M. Kamel, W. Hamouda, and A. Youssef, "Ultra-dense networks: A survey," *IEEE Communications Surveys Tutorials*, vol. 18, no. 4, pp. 2522–2545, 2016.

26. A. M. Abdalla, J. Rodriguez, I. Elfergani, and A. Teixeira, *Energy Efficiency in the Cloud Radio Access Network (CRAN) for 5G Mobile Networks*, 2019, pp. 225–248.

27. M. Li and H. Tsai, "Design and evaluation of a hybrid d2d discovery mechanism in 5g cellular networks," in *2018 Tenth International Conference on Ubiquitous and Future Networks (ICUFN)*, 2018, pp. 641–643.

28. A. Al-Fuqaha, M. Guizani, M. Mohammadi, M. Aledhari, and M. Ayyash, "Internet of things: A survey on enabling technologies, protocols, and applications," *IEEE Communications Surveys Tutorials*, vol. 17, no. 4, pp. 2347–2376, 2015.

29. F. Gao, W. Su, R. Shan, W. Zhu, K. He, and L. Wang, "3d coverage optimization research on 5g massive mimo antenna array," in *2018 IEEE International Symposium on Electromagnetic Compatibility and 2018 IEEE Asia-Pacific Symposium on Electromagnetic Compatibility (EMC/APEMC)*, 2018, pp. 94–97.

30. D. Lopez-Perez, I. Guvenc, G. de la Roche, M. Kountouris, T. Q. S. Quek, and J. Zhang, "Enhanced intercell interference coordination challenges in heterogeneous networks," *IEEE Wireless Communications*, vol. 18, no. 3, pp. 22–30, 2011.

31. H. Wang, S. Chen, M. Ai, and H. Xu, "Localized mobility management for 5g ultra dense network," *IEEE Transactions on Vehicular Technology*, vol. 66, no. 9, pp. 8535–8552, 2017.

32. L. Liu, V. Garcia, L. Tian, Z. Pan, and J. Shi, "Joint clustering and inter-cell resource allocation for comp in ultra dense cellular networks," in *2015 IEEE International Conference on Communications (ICC)*, 2015, pp. 2560–2564.

33. Y. Zhou, L. Liu, H. Du, L. Tian, X. Wang, and J. Shi, "An overview on intercell interference management in mobile cellular networks: From 2g to 5g," in *2014 IEEE International Conference on Communication Systems*, 2014, pp. 217–221.

34. M. Chen, Y. Hao, L. Hu, K. Huang, and V. K. N. Lau, "Green and mobility-aware caching in 5g networks," *IEEE Transactions on Wireless Communications*, vol. 16, no. 12, pp. 8347–8361, 2017.

35. H. Guo, J. Liu, and J. Zhang, "Computation offloading for multi-access mobile edge computing in ultra-dense networks," *IEEE Communications Magazine*, vol. 56, no. 8, pp. 14–19, 2018.

36. M. Jaber, M. A. Imran, R. Tafazolli, and A. Tukmanov, "5g backhaul challenges and emerging research directions: A survey," *IEEE Access*, vol. 4, pp. 1743–1766, 2016.

37. L. Li, M. Peng, Z. Yan, Z. Zhao, and Y. Li, "Success coverage probability for dynamic resource allocation in small cell networks," in *2016 IEEE Wireless Communications and Networking Conference*, 2016, pp. 1–5.

38. J. Ye, Y. He, X. Ge, and M. Chen, "Energy efficiency analysis of 5g ultra-dense networks based on random way point mobility models," in *2016 19th International Symposium on Wireless Personal Multimedia Communications (WPMC)*, 2016, pp. 177–182.

39. L. Su, C. Yang, and C. I, "On energy efficiency and spectral efficiency joint optimization of ultra dense networks," in *2015 IEEE Global Communications Conference (GLOBECOM)*, 2015, pp. 1–6.

40. Q. Ren, J. Fan, X. Luo, Z. Xu, and Y. Chen, "Analysis of spectral and energy efficiency in ultra-dense network," in *2015 IEEE International Conference on Communication Workshop (ICCW)*, 2015, pp. 2812–2817.

41. Y. Teng, M. Liu, F. R. Yu, V. C. M. Leung, M. Song, and Y. Zhang, "Resource allocation for ultra-dense networks: A survey, some research issues and challenges," *IEEE Communications Surveys Tutorials*, vol. 21, no. 3, pp. 2134–2168, 2019.

42. M. Haenggi, J. G. Andrews, F. Baccelli, O. Dousse, and M. Franceschetti, "Stochastic geometry and random graphs for the analysis and design of wireless networks," *IEEE Journal on Selected Areas in Communications*, vol. 27, no. 7, pp. 1029–1046, 2009.

43. R. Schneider and W. Weil, *Stochastic and integral geometry*. Springer, 2008.

44. M. Haenggi, *Stochastic Geometry for Wireless Networks*. Cambridge University Press, 2012.

45. J. G. Andrews, "Seven ways that hetnets are a cellular paradigm shift," *Communications Magazine IEEE*, vol. 51, no. 3, pp. 136–144, 2013.

46. J. G. Andrews, S. Buzzi, C. Wan, and S. V. Hanly, "What will 5g be?" *IEEE Journal on Selected Areas in Communications*, vol. 32, no. 6, pp. 1065–1082, 2014.

47. L. Zhang, H. C. Yang, and M. O. Hasna, "Generalized area spectral efficiency: An effective performance metric for green wireless communications," *IEEE Transactions on Communications*, vol. 62, no. 2, pp. 747–757, 2014.

48. E. Björnson, L. Sanguinetti, and M. Kountouris, "Deploying dense networks for maximal energy efficiency: Small cells meet massive mimo," *IEEE Journal on Selected Areas in Communications*, vol. 34, no. 4, pp. 832–847, 2015.

49. F. Baccelli, B. Blaszczyszyn, and P. Muhlethaler, "An aloha protocol for multihop mobile wireless networks," *IEEE Transactions on Information Theory*, vol. 52, no. 2, pp. 421–436, 2006.

50. R. Arshad, H. Elsawy, S. Sorour, T. Y. Al-Naffouri, and M. S. Alouini, "Handover management in dense cellular networks: A stochastic geometry approach," in *IEEE International Conference on Communications*, 2016.

51. T. Zhang, J. Zhao, L. An, and D. Liu, "Energy efficiency of base station deployment in ultra dense hetnets: A stochastic geometry analysis," *IEEE Wireless Communications Letters*, vol. 5, no. 2, pp. 184–187, 2016.

52. M. H. Manshaei, Q. Zhu, T. Alpcan, and J. P. Hubaux, "Game theory meets network security and privacy," *Acm Computing Surveys*, vol. 45, no. 3, pp. 1–39, 2013.

53. K. Binmore, "Does game theory work? : the bargaining challenge," *Mit Press Books*, vol. 1, 2007.

54. S. Samarakoon, M. Bennis, W. Saad, M. Debbah, and M. Latva-Aho, "Ultra dense small cell networks: Turning density into energy efficiency," *IEEE Journal on Selected Areas in Communications*, vol. 34, no. 5, pp. 1267–1280, 2016.

55. O. Guéant, J. M. Lasry, and P. L. Lions, *Mean Field Games and Applications*. Springer Berlin Heidelberg, 2011.

56. P. E. Caines, *Mean Field Games*. Springer London, 2015.

57. X. Mao, A. Maaref, and K. H. Teo, "Adaptive soft frequency reuse for inter-cell interference coordination in sc-fdma based 3gpp lte uplinks," in *IEEE GLOBECOM 2008 – 2008 IEEE Global Telecommunications Conference*, 2008, pp. 1–6.

58. L. Zhang, T. Jiang, and K. Luo, "Dynamic spectrum allocation for the downlink of ofdma-based hybrid-access cognitive femtocell networks," *IEEE Transactions on Vehicular Technology*, vol. 65, no. 3, pp. 1772–1781, 2016.

59. N. Michailow, M. Matthé, I. S. Gaspar, A. N. Caldevilla, L. L. Mendes, A. Festag, and G. Fettweis, "Generalized frequency division multiplexing for 5th generation cellular networks," *IEEE Transactions on Communications*, vol. 62, no. 9, pp. 3045–3061, 2014.

60. J. Zeng, T. Lv, R. P. Liu, X. Su, M. Peng, C. Wang, and J. Mei, "Investigation on evolving single-carrier noma into multi-carrier noma in 5g," *IEEE Access*, vol. 6, pp. 48 268–48 288, 2018.

61. S. M. R. Islam, N. Avazov, O. A. Dobre, and K. Kwak, "Power-domain non-orthogonal multiple access (noma) in 5g systems: Potentials and challenges," *IEEE Communications Surveys Tutorials*, vol. 19, no. 2, pp. 721–742, 2017.

62. T. Manglayev, R. C. Kizilirmak, Y. H. Kho, N. Bazhayev, and I. Lebedev, "Noma with imperfect sic implementation," in *IEEE EUROCON 2017 – 17th International Conference on Smart Technologies*, 2017, pp. 22–25.

63. N. M. Balasubramanya, A. Gupta, and M. Sellathurai, "Combining code-domain and power-domain noma for supporting higher number of users," in *2018 IEEE Global Communications Conference (GLOBECOM)*, 2018, pp. 1–6.

64. E. M. Eid, M. M. Fouda, A. S. T. Eldien, and M. M. Tantawy, "Performance analysis of musa with different spreading codes using ordered sic methods," in *2017 12th International Conference on Computer Engineering and Systems (ICCES)*, 2017, pp. 101–106.

65. H. Nikopour, E. Yi, A. Bayesteh, K. Au, M. Hawryluck, H. Baligh, and J. Ma, "Scma for downlink multiple access of 5g wireless networks," in *2014 IEEE Global Communications Conference*, 2014, pp. 3940–3945.

66. K. Vinod and R. Jose, "Compressed sensing algorithm for pattern division multiple access (pdma) in 5g radio networks," in *2018 International CET Conference on Control, Communication, and Computing (IC4)*, 2018, pp. 224–229.

67. Huawei, "5g: New air interface and radio access virtualization [eb/ol]," 2015.

68. M. You, X. Xin, W. Shangguang, L. Jinglin, S. Qibo, and Y. Fangchun, "Qos evaluation for web service recommendation," *China Communications*, vol. 12, no. 4, pp. 151–160, 2015.

69. J. Cui, Y. Liu, P. Fan, and A. Nallanathan, "A qoe-aware resource allocation strategy for multi-cell noma networks," in *2017 IEEE Globecom Workshops (GC Wkshps)*, 2017, pp. 1–6.

70. G. Taricco, "Optimum receiver design and performance analysis of arbitrarily correlated rician fading mimo channels with imperfect channel state information," *IEEE Transactions on Information Theory*, vol. 56, no. 3, pp. 1114–1134, 2010.

71. B. K. Chalise, Y. D. Zhang, and M. G. Amin, "Successive convex approximation for system performance optimization in a multiuser network with multiple mimo relays," in *2011 4th IEEE International Workshop on Computational Advances in Multi-Sensor Adaptive Processing (CAMSAP)*, 2011, pp. 229–232.

72. K. Janghel and S. Prakriya, "Performance of adaptive oma/cooperative-noma scheme with user selection," *IEEE Communications Letters*, vol. 22, no. 10, pp. 2092–2095, 2018.

73. L. Cao, X. Hu, M. Zhang, X. Wang, and X. Zhang, "Interactive comp with user-centric clustering based on load balancing in 5g dense networks," in *2018 IEEE International Conference on Communications Workshops (ICC Workshops)*, 2018, pp. 1–6.

74. J. Wu, J. Guo, E. W. M. Wong, and M. Zukerman, "Approximation of blocking probabilities in mobile cellular networks with channel borrowing," in *2015 IEEE 16th International Conference on High Performance Switching and Routing (HPSR)*, 2015, pp. 1–6.

75. P. Qiu, W. Chen, and J. Jiang, "Traffic offloading with mobility in lte henb networks," in *2015 IEEE 21st International Conference on Parallel and Distributed Systems (IC-PADS)*, 2015, pp. 809–814.

76. T. Toprasert and W. Lilakiataskun, "Tcp congestion control with mdp algorithm for iot over heterogeneous network," in *2017 17th International Symposium on Communications and Information Technologies (ISCIT)*, 2017, pp. 1–5.

77. M. Li, X. Xu, Y. Wang, and R. Zhang, "Game theory based load balancing in small cell heterogeneous networks," in *2015 International Conference on Connected Vehicles and Expo (ICCVE)*, 2015, pp. 26–31.

78. S. Nakazawa, N. Naganuma, and H. Otsuka, "Enhanced adaptive control cre in heterogeneous networks," in *2017 14th IEEE Annual Consumer Communications Networking Conference (CCNC)*, 2017, pp. 645–646.

79. D. Marabissi, G. Bartoli, and A. Stomaci, "Low-complexity distributed cell-specific bias calculation for load balancing in udns," *IEEE Transactions on Vehicular Technology*, vol. 68, no. 1, pp. 1056–1060, 2019.

80. H. Jiang, Z. Pan, N. Liu, X. You, and T. Deng, "Gibbs-sampling-based cre bias optimization algorithm for ultradense networks," *IEEE Transactions on Vehicular Technology*, vol. 66, no. 2, pp. 1334–1350, 2017.

81. J. Liu, Y. Yang, P. Sinha, and N. B. Shroff, "Load-adaptive base-station management for energy reduction including operation-cost and turn-on-cost," in *2017 IEEE Wireless Communications and Networking Conference (WCNC)*, 2017, pp. 1–6.

82. H. Liu, H. Cui, and J. Chen, "Dynamic sleeping algorithm of base station based on spatial features," in *2014 21st International Conference on Telecommunications (ICT)*, 2014, pp. 333–337.

83. A. Jushi, A. Pegatoquet, and T. N. Le, "Wind energy harvesting for autonomous wireless sensor networks," in *2016 Euromicro Conference on Digital System Design (DSD)*, 2016, pp. 301–308.

84. M. A. Marsan, G. Bucalo, A. Di Caro, M. Meo, and Y. Zhang, "Towards zero grid electricity networking: Powering bss with renewable energy sources," in *2013 IEEE International Conference on Communications Workshops (ICC)*, 2013, pp. 596–601.

85. P. N. Son and H. Y. Kong, "Cooperative communication with energy-harvesting relays under physical layer security," *IET Communications*, vol. 9, no. 17, pp. 2131–2139, 2015.

86. Y. Lv, H. Zhang, and S. Xueming, "Analysis of base stations deployment on power saving for heterogeneous network," in *2017 IEEE 17th International Conference on Communication Technology (ICCT)*, 2017, pp. 1439–1444.

87. H. Zhu, J. Mao, L. Wang, L. Fu, and N. Guo, "The study on point average energy consumption by monte carlo in large-scale wireless sensor networks," in *2015 IEEE International Conference on Information and Automation*, 2015, pp. 1700–1703.

88. A. He, D. Liu, Y. Chen, and T. Zhang, "Stochastic geometry analysis of energy efficiency in hetnets with combined comp and bs sleeping," in *2014 IEEE 25th Annual International Symposium on Personal, Indoor, and Mobile Radio Communication (PIMRC)*, 2014, pp. 1798–1802.

89. C. Liu, B. Natarajan, and H. Xia, "Small cell base station sleep strategies for energy efficiency," *IEEE Transactions on Vehicular Technology*, vol. 65, no. 3, pp. 1652–1661, 2016.

90. X. Xu, C. Yuan, W. Chen, X. Tao, and Y. Sun, "Adaptive cell zooming and sleeping for green heterogeneous ultradense networks," *IEEE Transactions on Vehicular Technology*, vol. 67, no. 2, pp. 1612–1621, 2018.

91. M. Zhao, J. Zhao, W. Zhou, J. Zhu, and S. Zhang, "Energy efficiency optimization in relay-assisted networks with energy harvesting relay constraints," *China Communications*, vol. 12, no. 2, pp. 84–94, 2015.

92. A. Bozorgchenani, D. Tarchi, and G. E. Corazza, "Centralized and distributed architectures for energy and delay efficient fog network-based edge computing services," *IEEE Transactions on Green Communications and Networking*, vol. 3, no. 1, pp. 250–263, 2019.

93. T. Wu, X. Li, H. Ji, and H. Zhang, "An energy-efficient sleep management algorithm for udn with edge caching," in *2017 IEEE Globecom Workshops (GC Wkshps)*, 2017, pp. 1–5.

94. B. Liu, M. Zhao, W. Zhou, J. Zhu, and P. Dong, "Flow-level-delay constraint small cell sleeping with macro base station cooperation for energy saving in hetnet," in *2015 IEEE 82nd Vehicular Technology Conference (VTC2015-Fall)*, 2015, pp. 1–5.

95. J. Yang, X. Zhang, and W. Wang, "Two-stage base station sleeping scheme for green cellular networks," *Journal of Communications and Networks*, vol. 18, no. 4, pp. 600–609, 2016.

96. J. Liu, M. Sheng, L. Liu, and J. Li, "Interference management in ultra-dense networks: Challenges and approaches," *IEEE Network*, vol. 31, no. 6, pp. 70–77, 2017.

97. J. L. W. Nam, D. Bai, and I. Kang, "Advanced interference management for 5g cellular networks," *IEEE Communications Magazine*, vol. 52, pp. 52–60, 2014.

98. L. J. Z. Haq Abbas, F. Muhammad, "Analysis of load balancing and interference management in heterogeneous cellular networks," *IEEE Access*, vol. 5, pp. 14 690–14 705, 2017.

99. L. B. L. N. Saquib, E. Hossain, and D. I. Kim, "Interference management in ofdma femtocell networks: issues and approachess," *IEEE Wireless Communications*, vol. 19, pp. 86–95, 2012.

100. M. I. Kamel and K. M. F. Elsayed, "Performance evaluation of a coordinated time-domain eicic framework based on absf in heterogeneous lte-advanced networks," in *2012 IEEE Global Communications Conference (GLOBECOM)*, 2012, pp. 5326–5331.

101. G. Fodor, C. Koutsimanis, A. Rácz, N. Reider, A. Simonsson, and W. Müller, "Intercell interference coordination in ofdma networks and in the 3gpp long term evolution system," *Journal of Communications*, vol. 4, pp. 445–453, 2009.

102. G. RP-091440, "Work item description: Carrier aggregation for LTE," 2009.

103. S. C. S. Park, W. Seo, and D. Hong., "A beamforming codebook restriction for cross-tier interference coordination in two-tier femtocell networks," *IEEE Transactions on Vehicular Technology*, vol. 60, pp. 1651–1663, 2011.

104. P. Mogensen, K. Pajukoski, E. Tiirola, J. Vihriala, E. Lahetkangas, G. Berardinelli, F. M. L. Tavares, N. H. Mahmood, M. Lauridsen, D. Catania, and A. F. Cattoni, "Centimeter-wave concept for 5g ultra-dense small cells," in *2014 IEEE 79th Vehicular Technology Conference (VTC Spring)*, 2014, pp. 1–6.

105. N. T. J. B. Soret, K. I. Pedersen, and V. Fernández-López., "Interference coordination for dense wireless networks," *IEEE Communications Magazine*, vol. 53, pp. 102–109, 2015.

106. S. K. R. Jain and N. Agrawal, "Smart antenna for cellular mobile communication," *arXiv preprint arXiv*, 2012.

107. K. Yao, Q. Wu, Y. Xu, and J. Jing, "Distributed abs-slot access in dense heterogeneous networks: A potential game approach with generalized interference model," *IEEE Access*, vol. 5, pp. 94–104, 2017.

108. J. Xu, Y. Wang, Q. Zhang, L. Xu, and S. Wu, "Fairness-aware interference coordination by combined sfr and comp for heterogeneous networks," *2017 IEEE 86th Vehicular Technology Conference*, 2017.

109. G. boggia, M. Kurras, W. Zirwas, R. SivaSiva Ganesan, K. Pedersen, L. Thiele, A. Grassi, and G. Piro, "Massive mimo interference coordination for 5g broadband access: Integration and system level study," *Computer Networks*, vol. 147, pp. 191–203, 2018.

110. H. Ning, H. Liu, J. Ma, L. T. Yang, Y. Wan, X. Ye, and R. Huang, "From internet to smart world," *IEEE Access*, vol. 3, pp. 1994–1999, 2015.

111. N. Ul Hasan, W. Ejaz, N. Ejaz, H. S. Kim, A. Anpalagan, and M. Jo, "Network selection and channel allocation for spectrum sharing in 5g heterogeneous networks," *IEEE Access*, vol. 4, pp. 980–992, 2016.

112. E. Z. Tragos, S. Zeadally, A. G. Fragkiadakis, and V. A. Siris, "Spectrum assignment in cognitive radio networks: A comprehensive survey," *IEEE Communications Surveys Tutorials*, vol. 15, no. 3, pp. 1108–1135, 2013.

113. S. Haykin, "Cognitive radio: brain-empowered wireless communications," *IEEE Journal on Selected Areas in Communications*, vol. 23, no. 2, pp. 201–220, 2005.

114. R. H. Tehrani, S. Vahid, D. Triantafyllopoulou, H. Lee, and K. Moessner, "Licensed spectrum sharing schemes for mobile operators: A survey and outlook," *IEEE Communications Surveys Tutorials*, vol. 18, no. 4, pp. 2591–2623, 2016.

115. G. P. Koudouridis and P. Soldati, "Joint network density and spectrum sharing in multi-operator collocated ultra-dense networks," in *2018 7th International Conference on Modern Circuits and Systems Technologies (MOCAST)*, 2018, pp. 1–4.

116. J. Park, S. Kim, and J. Zander, "Asymptotic behavior of ultra-dense cellular networks and its economic impact," in *2014 IEEE Global Communications Conference*, 2014, pp. 4941–4946.

117. C. Fan, B. Li, C. Zhao, W. Guo and Y. -C. Liang, "Learning-based spectrum sharing and spatial reuse in mm-wave ultradense networks," *IEEE Transactions on Vehicular Technology*, vol. 67, pp. 4954–498, 2018.

118. C. G. S. Chen, and Z. Zeng, "Polarization-based multi-dimensional resource optimization for spectrum sharing in dense heterogeneous networks," *2018 IEEE International Conference on Communications Workshops*, 2018.

119. F. Molisch, V. Ratnam, S. Han, and Z. Li, "Hybrid beamforming for massive mimo: A survey," *IEEE Communications Magazine*, vol. 55, pp. 134–141, 2017.

120. L. Lu, Y. Li, A. Lee, A. Ashikhmin, and R. Zhang, "An overview of massive mimo: Benefits and challenges," *IEEE Journal of Selected Topics in Signal Processing*, vol. 8, pp. 742–758, 2014.

121. S. Han, C. I, Z. Xu, and C. Rowell, "Large-scale antenna systems with hybrid analog and digital beamforming for millimeter wave 5g," *IEEE Communications Magazine*, vol. 53, no. 1, pp. 186–194, 2015.

122. S. Kutty and D. Sen, "Beamforming for millimeter wave communications: An inclusive survey," *IEEE Communications Surveys and Tutorials*, vol. 18, pp. 949–973, 2016.

123. F. S. W. Yu, "Hybrid digital and analog beamforming design for large-scale antenna arrays," *IEEE Journal of Selected Topics in Signal Processing*, vol. 10, pp. 501–513, 2016.

124. J. Turkka, M. Koivisto, J. Werner, A. Hakkarainen, P. Kela, M. Valkama, R. Jantti, K. Leppanen, and M. Costa, "Location based beamforming in 5g ultra-dense networks," *2016 IEEE 84th Vehicular Technology Conference*, 2016.

125. C. Xing, N. Wang, J. Ni, Z. Fei, and J. Kuang, "Mimo beamforming designs with partial csi under energy harvesting constraints," *IEEE Signal Processing Letters*, vol. 20, pp. 363–366, 2013.

126. M. Kamel, W. Hamouda, and A. Youssef, "Ultra-dense networks: A survey," *IEEE Communications Surveys Tutorials*, vol. 18, no. 4, pp. 2522–2545, 2016.

127. H. Tabassum, U. Siddique, E. Hossain, and M. J. Hossain, "Downlink performance of cellular systems with base station sleeping, user association, and scheduling," *IEEE Network Transactions on Wireless Communications*, vol. 13, no. 10, pp. 5752–5767, 2014.

128. D. L. Perez, X. Chu, and I. Guvenc, "On the expanded region of picocells in heterogeneous networks," *IEEE Journal of Selected Topics in Signal Processing.*, vol. 6, no. 3, pp. 281–294, 2012.

129. N. Himayat, S. Talwar, and A. Rao, "Interference management for 4g cellular standards [WIMAX/LTE update]," *Communications Magazine IEEE*, vol. 48, no. 8, pp. 86–92, 2010.

130. S. Lahoud, K. Khawam, and S. Martin, "Energy efficient joint scheduling and power control in multi-cell wireless networks," *IEEE Journal on Selected Areas in Communications*, vol. 99, no. 1, pp. 1–1, 2016.

131. S. Bassoy, H. Farooq, and M. Imran, "Coordinated multi-point clustering schemes: A survey," *IEEE Communications Surveys and Tutorials*, vol. 1, no. 1, pp. 1–1, 2017.

132. X. Ge, H. Jin, and J. Cheng, "On fair resource sharing in downlink coordinated multi-point systems," *IEEE Communications Letters*, vol. 20, no. 6, pp. 1–1, 2016.

133. H. Elsawy, E. Hossain, and M. Alouini, "Analytical modeling of mode selection and power control for underlay d2d communication in cellular networks," *IEEE Transactions on Communications*, vol. 62, no. 11, pp. 4147–4161, 2014.

134. N. Lee, X. Lin, and J. G. Andrews, "Power control for d2d underlaid cellular networks: Modeling, algorithms, and analysis," *IEEE Journal on Selected Areas in Communications*, vol. 33, no. 1, pp. 1–13, 2015.

135. W. Mei and R. Zhang, "Uplink cooperative noma for cellular-connected uav," *IEEE Journal of Selected Topics in Signal Processing*, vol. 13, no. 3, pp. 1–1, 2019.

136. D. Hui, H. Zhang, and T. Jie, "Energy efficient user association and power control for dense heterogeneous networks," in *International Conference on Computing*, 2018.

137. R. Aslani and M. Rasti, "Distributed power control schemes in in-band full-duplex energy harvesting wireless networks," *IEEE Transactions on Wireless Communications*, vol. 9, no. 9, pp. 1–1, 2018.

138. Wikipedia, "Handover," in *http://wikipedia.moesalih.com/*, 2019.

139. E. R. Bastidas-Puga, G. Galaviz, and D. H. Covarrubias, "Handover based on a predictive approach of signal-to-interference-plus-noise ratio for heterogeneous cellular networks," *IET Communications*, vol. 13, no. 6, pp. 672–678, 2019.

140. J. Yang, X. Ji, and K. Huang, "Unified and fast handover authentication based on link signatures in 5g sdn-based hetnet," *IET Communications*, vol. 13, no. 2, pp. 144–152, 2019.

141. T. Bilen, B. Canberk, and K. R. Chowdhury, "Handover management in software-defined ultra-dense 5g networks," *IEEE Network*, vol. 31, no. 4, pp. 49–55, 2017.

142. R. N. Clarke, "Expanding mobile wireless capacity: The challenges presented by technology and economics ," *Telecommunications Policy*, vol. 38, no. 8-9, pp. 693–708, 2014.

143. H. Q. Ngo, "Massive mimo: Fundamentals and system designs." *Linköping University Electronic Press*, vol. 1642, 2015.

144. T. L. Marzetta, "Noncooperative cellular wireless with unlimited numbers of base station antennas," *IEEE Transactions on Wireless Communications*, vol. 9, no. 11, pp. 3590–3600, 2010.

145. E. G. Larsson, O. Edfors, F. Tufvesson, and T. L. Marzetta, "Massive mimo for next generation wireless systems," *IEEE Communications Magazine*, vol. 52, no. 2, pp. 186–195, 2014.

146. J. Jose, A. Ashikhmin, T. L. Marzetta, and S. Vishwanath, "Pilot contamination and precoding in multi-cell tdd systems," *IEEE Transactions on Wireless Communications*, vol. 10, no. 8, pp. 2640–2651, 2011.

147. B. Gopalakrishnan and N. Jindal, "An analysis of pilot contamination on multi-user mimo cellular systems with many antennas," in *IEEE International Workshop on Signal Processing Advances in Wireless Communications*, 2011.

148. F. Fernandes, A. Ashikhmin, and T. L. Marzetta, "Inter-cell interference in noncooperative tdd large scale antenna systems," *IEEE Journal on Selected Areas in Communications*, vol. 31, no. 2, pp. 192–201, 2013.

149. L. An and V. Lau, "Hierarchical interference mitigation for massive mimo cellular networks," *IEEE Transactions on Signal Processing*, vol. 62, no. 18, pp. 4786–4797, 2014.

150. H. Ali-Ahmad, C. Cicconetti, A. D. L. Oliva, M. Draxler, R. Gupta, V. Mancuso, L. Roullet, and V. Sciancalepore, "Crowd: An sdn approach for densenets," in *Second European Workshop on Software Defined Networks*, 2013, pp. 25–31.

151. J. Nam, A. Adhikary, J. Y. Ahn, and G. Caire, "Joint spatial division and multiplexing: Opportunistic beamforming, user grouping and simplified downlink scheduling," *IEEE Journal of Selected Topics in Signal Processing*, vol. 8, no. 5, pp. 876–890, 2017.

152. L. An and V. K. N. Lau, "Phase only rf precoding for massive mimo systems with limited rf chains," *IEEE Transactions on Signal Processing*, vol. 62, no. 17, pp. 4505–4515, 2014.

153. T. E. Bogale, L. B. Le, A. Haghighat, and L. Vandendorpe, "On the number of rf chains and phase shifters, and scheduling design with hybrid analog-digital beamforming," *IEEE Transactions on Wireless Communications*, vol. 15, no. 5, pp. 3311–3326, 2016.

154. P. Sudarshan, N. B. Mehta, A. F. Molisch, and J. Zhang, "Channel statistics-based rf pre-processing with antenna selection," *IEEE Transactions on Wireless Communications*, vol. 5, no. 12, pp. 3501–3511, 2006.

155. A. Alkhateeb, O. E. Ayach, G. Leus, and R. W. Health Jr, "Channel estimation and hybrid precoding for millimeter wave cellular systems," *IEEE Journal of Selected Topics in Signal Processing*, vol. 8, no. 5, pp. 831–846, 2017.

156. R. Taori and A. Sridharan, "Point-to-multipoint in-band mmwave backhaul for 5g networks," *IEEE Communications Magazine*, vol. 53, no. 1, pp. 195–201, 2015.

157. Z. Gao, L. Dai, D. Mi, and Z. Wang, "Mmwave massive-mimo-based wireless backhaul for the 5g ultra-dense network," *Wireless Communications IEEE*, vol. 22, no. 5, pp. 13–21, 2015.

158. C. Fan, Y.-J. A. Zhang, and X. Yuan, "Machine learning for heterogeneous ultra-dense networks with graphical representations," *arXiv preprint arXiv:1808.04547*, 2018.

159. A. L. Samuel, "Some studies in machine learning using the game of checkers," *IBM Journal of Research and Development*, vol. 3, no. 3, pp. 210–229, 1959.

160. Z. Fadlullah, F. Tang, B. Mao, N. Kato, O. Akashi, T. Inoue, and K. Mizutani, "State-of-the-art deep learning: Evolving machine intelligence toward tomorrows intelligent network traffic control systems," *IEEE Communications Surveys & Tutorials*, vol. 19, no. 4, pp. 2432–2455, 2017.

161. C. Jiang, H. Zhang, Y. Ren, Z. Han, K.-C. Chen, and L. Hanzo, "Machine learning paradigms for next-generation wireless networks," *IEEE Wireless Communications*, vol. 24, no. 2, pp. 98–105, 2017.

162. B. K. Donohoo, C. Ohlsen, S. Pasricha, Y. Xiang, and C. Anderson, "Context-aware energy enhancements for smart mobile devices," *IEEE Transactions on Mobile Computing*, vol. 13, no. 8, pp. 1720–1732, 2014.

163. C. K. Wen, S. Jin, K. K. Wong, J. C. Chen, and P. Ting, "Channel estimation for massive mimo using gaussian-mixture bayesian learning," *IEEE Transactions on Wireless Communications*, vol. 14, no. 3, pp. 1356–1368, 2015.

164. K. W. Choi and E. Hossain, "Estimation of primary user parameters in cognitive radio systems via hidden markov model," *IEEE Transactions on Signal Processing*, vol. 61, no. 3, pp. 782–795, 2013.

165. H. Harai, M. Inoue, M. Xia, and Y. Owada, "Optical and wireless hybrid access networks: Design and optimization," *IEEE/OSA Journal of Optical Communications & Networking*, vol. 4, no. 10, pp. 749–759, 2012.

166. R. C. Qiu, Z. Hu, Z. Chen, N. Guo, R. Ranganathan, S. Hou, and G. Zheng, "Cognitive radio network for the smart grid: Experimental system architecture, control algorithms, security, and microgrid testbed," *IEEE Transactions on Smart Grid*, vol. 2, no. 4, pp. 724–740, 2011.

167. A. Aprem, C. R. Murthy, and N. B. Mehta, "Transmit power control policies for energy harvesting sensors with retransmissions," *IEEE Journal of Selected Topics in Signal Processing*, vol. 7, no. 5, pp. 895–906, 2013.

168. O. Onireti, A. Zoha, J. Moysen, A. Imran, L. Giupponi, M. A. Imran, and A. Abu-Dayya, "A cell outage management framework for dense heterogeneous networks," *IEEE Transactions on Vehicular Technology*, vol. 65, no. 4, pp. 2097–2113, 2016.

169. S. Maghsudi and S. Stanczak, "Channel selection for network-assisted d2d communication via no-regret bandit learning with calibrated forecasting," *IEEE Transactions on Wireless Communications*, vol. 14, no. 3, pp. 1309–1322, 2015.

170. G. Shen, L. Pei, P. Zhiwen, L. Nan, and Y. Xiaohu, "Machine learning based small cell cache strategy for ultra dense networks," in *Wireless Communications and Signal Processing (WCSP), 2017 9th International Conference on*. IEEE, 2017, pp. 1–6.

171. V. S. Feng and S. Y. Chang, "Determination of wireless networks parameters through parallel hierarchical support vector machines," *IEEE Transactions on Parallel & Distributed Systems*, vol. 23, no. 3, pp. 505–512, 2012.

172. A. Assra, J. Yang, and B. Champagne, "An em approach for cooperative spectrum sensing in multi-antenna CR networks," *IEEE Transactions on Vehicular Technology*, vol. 65, no. 3, pp. 1229–1243, 2016.

173. C. K. Yu, K. C. Chen, and S. M. Cheng, "Cognitive radio network tomography," *IEEE Transactions on Vehicular Technology*, vol. 59, no. 4, pp. 1980–1997, 2010.

174. H. Nguyen, G. Zheng, H. Zhu, and R. Zheng, "Binary inference for primary user separation in cognitive radio networks," *IEEE Transactions on Wireless Communications*, vol. 12, no. 4, pp. 1532–1542, 2013.

175. G. Alnwaimi, S. Vahid, and K. Moessner, "Dynamic heterogeneous learning games for opportunistic access in lte-based macro/femtocell deployments," *IEEE Transactions on Wireless Communications*, vol. 14, no. 4, pp. 2294–2308, 2015.

176. H. Farooq, M. S. Parwez, and A. Imran, "Continuous time markov chain based reliability analysis for future cellular networks," in *IEEE Global Communications Conference*, 2017, pp. 1–6.

177. B. N. Astuto, M. M. Ca, N. N. Xuan, K. Obraczka, and T. Turletti, "A survey of software-defined networking: Past, present, and future of programmable networks," *IEEE Communications Surveys & Tutorials*, vol. 16, no. 3, pp. 1617–1634, 2014.

178. E. Summary, "Software-defined networking: The new norm for networks," 2015.

179. D. Kreutz, F. M. V. Ramos, P. Verissimo, C. E. Rothenberg, S. Azodolmolky, and S. Uhlig, "Software-defined networking: A comprehensive survey," *Proceedings of the IEEE*, vol. 103, no. 1, pp. 10–13, 2014.

180. V. C. M. Borges, K. V. Cardoso, E. Cerqueira, M. Nogueira, and A. Santos, "Aspirations, challenges, and open issues for software-based 5g networks in extremely dense and heterogeneous scenarios," *Eurasip Journal on Wireless Communications & Networking*, vol. 2015, no. 1, pp. 1–13, 2015.

181. M. Hadzialic, B. Dosenovic, M. Dzaferagic, and J. Musovic, "Cloud-ran: Innovative radio access network architecture," in *Elmar, 2013 International Symposium*, 2013, pp. 115–120.

182. A. Checko, H. L. Christiansen, Y. Yan, L. Scolari, G. Kardaras, M. S. Berger, and L. Dittmann, "Cloud ran for mobile networksa technology overview," *IEEE Communications Surveys & Tutorials*, vol. 17, no. 1, pp. 405–426, 2014.

183. J. Wu, Z. Zhang, Y. Hong, and Y. Wen, "Cloud radio access network (c-ran): a primer," *Network IEEE*, vol. 29, no. 1, pp. 35–41, 2015.

184. Y. Teng and W. Zhao, "Robust group sparse beamforming for dense c-rans with probabilistic sinr constraints," in *Wireless Communications & Networking Conference*, 2017.

185. B. Dai and Y. Wei, "Sparse beamforming and user-centric clustering for downlink cloud radio access network," *IEEE Access*, vol. 2, pp. 1326–1339, 2017.

186. C. Liu, K. Sundaresan, M. Jiang, S. Rangarajan, and G. K. Chang, "The case for reconfigurable backhaul in cloud-RAN based small cell networks," in *IEEE INFOCOM*, 2013, pp. 1124–1132.

187. Z. Pi and F. Khan, "An introduction to millimeter-wave mobile broadband systems," *Communications Magazine IEEE*, vol. 49, no. 6, pp. 101–107, 2011.

188. M. Elkashlan, T. Q. Duong, and H. H. Chen, "Millimeter-wave communications for 5g: Fundamentals: Part i (guest editorial)," *Communications Magazine IEEE*, vol. 52, no. 9, pp. 52–54, 2014.

189. N. Yong, L. Yong, D. Jin, S. Li, and A. V. Vasilakos, "A survey of millimeter wave communications (mmwave) for 5g: opportunities and challenges," *Wireless Networks*, vol. 21, no. 8, pp. 2657–2676, 2015.

190. Q. Chen, R. Y. Fei, H. Tao, R. Xie, L. Jiang, and Y. Liu, "An integrated framework for software defined networking, caching and computing," *IEEE Network*, vol. 31, no. 3, pp. 12–21, 2017.

191. T. O. Olwal, K. Djouani, and A. M. Kurien, "A survey of resource management towards 5g radio access networks," *IEEE Communications Surveys & Tutorials*, vol. 18, no. 3, pp. 1656–1686, 2016.

192. N. Lu, N. Cheng, N. Zhang, X. Shen, and J. W. Mark, "Connected vehicles: Solutions and challenges," *Internet of Things Journal IEEE*, vol. 1, no. 4, pp. 289–299, 2014.

193. P. Bedi and V. Jindal, "Use of big data technology in vehicular ad-hoc networks," in *International Conference on Advances in Computing*, 2014.

194. C. Nan, L. Feng, J. Chen, W. Xu, and Shen, "Big data driven vehicular networks," *IEEE Network*, vol. PP, no. 99, pp. 1–8, 2018.

195. R. Baldemair, T. Irnich, K. Balachandran, E. Dahlman, G. Mildh, Y. Selén, S. Parkvall, M. Meyer, and A. Osseiran, "Ultra-dense networks in millimeter-wave frequencies," *Communications Magazine IEEE*, vol. 53, no. 1, pp. 202–208, 2015.

196. M. Gregori, J. Gómez-Vilardebó, J. Matamoros, and D. Gündüz, "Wireless content caching for small cell and d2d networks," *IEEE Journal on Selected Areas in Communications*, vol. 34, no. 5, pp. 1222–1234, 2016.

197. R. Vannithamby and S. Talwar, "Proactive caching in 5g small cell networks," *Diesel Progress North American ...*, no. March, 2016.

198. Y. Yue and J. Zhu, "Write skew and zipf distribution: Evidence and implications," *ACM Transactions on Storage*, vol. 12, no. 4, p. 21, 2016.

199. N. Golrezaei, A. F. Molisch, A. G. Dimakis, and G. Caire, "Femtocaching and device-to-device collaboration: A new architecture for wireless video distribution," *IEEE Communications Magazine*, vol. 51, no. 4, pp. 142–149, 2013.

200. S. Yue, M. Larson, and A. Hanjalic, "Collaborative filtering beyond the user-item matrix:a survey of the state of the art and future challenges," *ACM Computing Surveys*, vol. 47, no. 1, pp. 1–45, 2014.

201. E. Bastug, M. Bennis, and M. Debbah, "Living on the edge: The role of proactive caching in 5g wireless networks," *IEEE Communications Magazine*, vol. 52, no. 8, pp. 82–89, 2014.

202. L. Atzori, A. Iera, and G. Morabito, "The internet of things: A survey," *Computer Networks*, vol. 54, no. 15, pp. 2787–2805, 2010.

203. A. Al-Fuqaha, M. Guizani, M. Mohammadi, M. Aledhari, and M. Ayyash, "Internet of things: A survey on enabling technologies, protocols and applications," *IEEE Communications Surveys & Tutorials*, vol. 17, no. 4, pp. 2347–2376, 2015.

204. S. Severi, F. Sottile, G. Abreu, C. Pastrone, M. Spirito, and F. Berens, "M2M technologies: Enablers for a pervasive internet of things," in *European Conference on Networks & Communications*, 2014.

205. M. E. Tanab, W. Hamouda, M. E. Tanab, and W. Hamouda, "A scalable overload control algorithm for massive access in machine-to-machine networks," in *IEEE International Conference on Communications*, 2017.

206. Z. Dawy, W. Saad, A. Ghosh, J. G. Andrews, and E. Yaacoub, "Toward massive machine type cellular communications," *IEEE Wireless Communications*, vol. 24, no. 1, pp. 120–128, 2017.

207. B. Panigrahi, H. K. Rath, R. Ramamohan, and A. Simha, "Energy and spectral efficient direct machine-to-machine (m2m) communication for cellular internet of things (iot) networks," in *International Conference on Internet of Things & Applications*, 2016.

208. A. Darif, "Ultra low power consumption, short transmission time and efficient data coding wireless communication technology for mwsns," in *International Conference on Multimedia Computing & Systems*, 2016.

209. M. I. Hossain, A. Azari, and J. Zander, "Dera: Augmented random access for cellular networks with dense h2h-mtc mixed traffic," in *Globecom Workshops*, 2017.

210. C. Zheng, N. Pappas, M. Kountouris, and V. Angelakis, "Throughput analysis of smart objects with delay constraints," in *IEEE International Symposium on A World of Wireless*, 2016.

211. S. Sardellitti, G. Scutari, and S. Barbarossa, "Joint optimization of radio and computational resources for multicell mobile-edge computing," in *IEEE International Workshop on Signal Processing Advances in Wireless Communications*, 2014.

212. Y. Mao, J. Zhang, and K. B. Letaief, "Joint task offloading scheduling and transmit power allocation for mobile-edge computing systems," in *Wireless Communications & Networking Conference*, 2017.

213. C. Xu, L. Pu, G. Lin, W. Wu, and W. Di, "Exploiting massive d2d collaboration for energy-efficient mobile edge computing," *IEEE Wireless Communications*, vol. 24, no. 4, pp. 64–71, 2017.

214. H. Zhen, Z. Wang, C. Wei, and V. C. M. Leung, "Blockchain-empowered fair computational resource sharing system in the d2d network," *Future Internet*, vol. 9, no. 4, p. 85, 2017.

215. H. Ru, R. Y. Fei, H. Tao, R. Xie, L. Jiang, V. C. M. Leung, and Y. Liu, "Software defined networking, caching, and computing for green wireless networks," *IEEE Communications Magazine*, vol. 54, no. 11, pp. 185–193, 2016.

216. Z. Nan, X. Liu, F. R. Yu, L. Ming, and V. C. M. Leung, "Communications, caching, and computing (3C)-oriented small-cell networks with interference alignment (IA)," *IEEE Communications Magazine*, vol. 54, no. 9, pp. 29–35, 2016.

217. S. Ulukus, A. Yener, E. Erkip, O. Simeone, M. Zorzi, P. Grover, and K. Huang, "Energy harvesting wireless communications: A review of recent advances," *IEEE Journal on Selected Areas in Communications*, vol. 33, no. 3, pp. 360–381, 2015.

218. S. Rostami, K. Heiska, O. Puchko, G. P. Koudouridis, K. Leppanen, and M. Valkama, "Wireless backhauling for energy harvesting ultra-dense networks," in *2018 IEEE 29th Annual International Symposium on Personal, Indoor and Mobile Radio Communications (PIMRC)*, 2018, pp. 1807–1812.

219. S. Al-Sultan, M. M. Al-Doori, A. H. Al-Bayatti, and H. Zedan, "A comprehensive survey on vehicular ad hoc network," *Journal of Network & Computer Applications*, vol. 37, no. 1, pp. 380–392, 2014.

Index